Collins

D0413183

Collins Revision KS3 Maths Revision Guide Levels 5–8

Keith Gordon

800 094 872

Revision contents

NUMBER

ALGEBRA

SHAPE, SPACE AND MEASURES

HANDLING DATA

Workbook contents

NUMBER

Multiplying and dividing decimals

level 5

Multiplying by powers of 10

- When you **multiply** by **10** all the digits move **one** place to the **left**.
- When you **multiply** by **10^2** all the digits move **two** places to the **left**.

Example: Work out **a** 2.79 x 10 **b** 3.2×10^2

a Tens Units . Tenths Hundredths

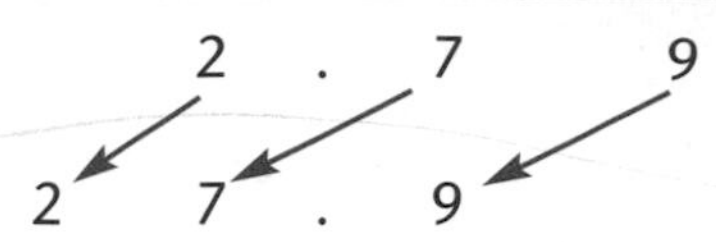

b Hundreds Tens Units . Tenths Hundredths

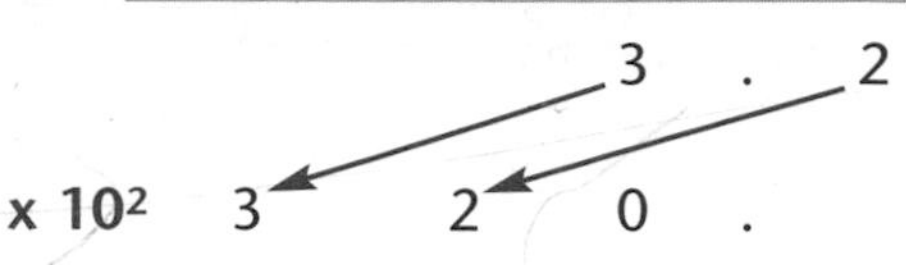

- You will need to add a zero in part **b** and you do not have to put in the decimal point for a whole number.

Top Tip!

When multiplying by 10 the digits move one place to the left but you can also move the decimal point one place to the right.

level 5

Dividing by powers of 10

- When you **divide** by **10** all the digits move **one** place to the **right**.
- When you **divide** by **10^2** all the digits move **two** places to the **right**.

Example: Work out **a** 32.4 ÷ 10 **b** $2.7 \div 10^2$

a Tens Units . Tenths Hundredths

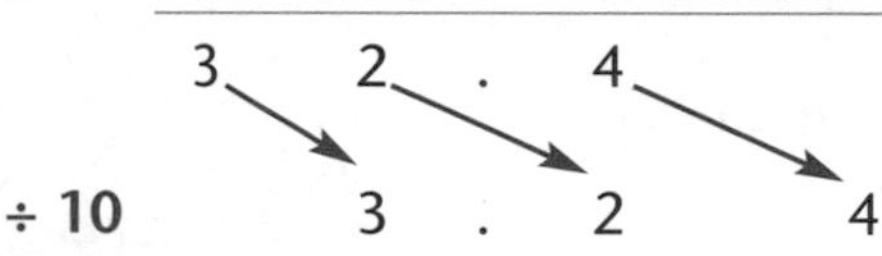

b Units . Tenths Hundredths Thousands

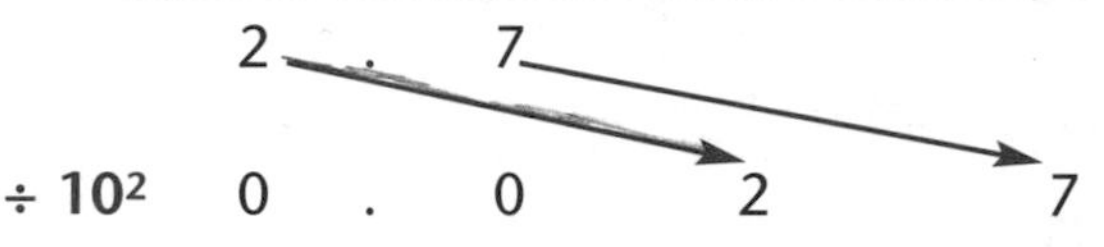

- You will need to add a zero before and after the decimal point in part **b**.

Top Tip!

When dividing by 10 the digits move one place to the right but you can also move the decimal point one place to the left.

level 5

Multiplying and dividing decimals

- Multiplying and dividing decimals is just like normal multiplying and dividing. All you have to do is keep the decimal point lined up.
- The decimal point in the answer is always underneath the decimal point in the question.

Example: Work out 3.76 x 4

```
   3 . 7 6
x        4
----------
 1 5 . 0 4
   3   2
```

Example: Work out 8.04 ÷ 6

```
    1 . 3 4
6 | 8 .²0 ²4
```

Top Tip!

Estimate the answer just to be sure.
3.76 x 4 is approximately 4 x 4 = 16
8.04 ÷ 6 is approximately 9 ÷ 6 = 1.5

Sample mental test questions

A book costs £2.99. How much do 5 books cost?

5 lots of £3 are £15. £15 – 5p = £14.95

Divide 3.45 by 3.

Split the calculation:

$(3 \div 3) + (0.45 \div 3) = 1 + 0.15 = 1.15$

level 5

Sample worked test question

a A shop sells boxes of chocolates for £3.38.
How much do 10 boxes cost?

b A box of 6 Easter eggs costs £16.20.
How much does each egg cost?

Answers

a *Move the digits one place to the left: 3.38 x 10 = 33.8*
10 boxes will cost £33.80.

b
$$\begin{array}{r} 2\,.\,7\,0 \\ 6\overline{)\,1\,6\,.^{4}2\,0} \end{array}$$
Each egg costs £2.70.

Top Tip!
Don't forget to put a zero on the end as the answer is in pounds and pence.

Top Tip!
Don't forget to estimate:
$16.20 \div 6 \approx 18 \div 6 = 3$
'Is approximately equal to' can be shown by this symbol: $\approx$

Did You Know?
The world's most expensive Easter egg is the $24 million Fabergé Coronation Egg.

Spot Check **1** Work out **a** 0.7×10^2 **b** $4.52 \div 10^3$ **c** 3.12×6 **d** $7.35 \div 5$

NUMBER

Long multiplication and division

level 5

Long multiplication

- There are many ways of doing long multiplication. Two are shown here.

The **standard column** method.

Example: Work out 32 x 256

```
    2 5 6
x     3 2
    -----
    5 1 2
+ 7 6 8 0
  -------
  8 1 9 2
```

The **box** method.

Example: Work out 43 x 264

X	200	60	4
40	8000	2400	160
3	600	180	12

Add up the numbers in the boxes.

```
              8 0 0 0
              2 4 0 0
                1 6 0
                6 0 0
                1 8 0
+                 1 2
            ---------
43 x 264 =  1 1 3 5 2
```

Top Tip!

Decide which method you prefer and stick with it.

level 5

Long division

- There are two ways of doing long division.

The **standard column** method.

Example: Work out 962 ÷ 37

```
       2 6
37 | 9 6 2
   - 7 4
   -------
     2 2 2
   - 2 2 2
   -------
         0
```

The repeated subtraction or '**chunking**' method.

Example: Work out 896 ÷ 28

```
  8 9 6
- 5 6 0    20 x 28
  -----
  3 3 6
- 2 8 0    10 x 28
  -----
    5 6
-   5 6     2 x 28
  -----
      0    32 x 28
```

So, 896 ÷ 28 = 20 + 10 + 2 = 32

Top Tip!

Some test questions usually ask part **a** as a long multiplication and part **b** as a long division and put the question into a real-life situation.

Top Tip!

Write out some of the easier times tables for the divisor:

1 x 28 = 28
2 x 28 = 56
4 x 28 = 112
10 x 28 = 280
20 x 28 = 560

Subtract the biggest multiple you can each time.

level 5

Sample worked test question

a A garden centre has 576 winter pansies for sale.
Each plant costs 28p.
How much will all 576 cost?

b The pansies are packed in trays of 18.
How many trays does the garden centre have?

Answers

a *Using the box method:*

X	500	70	6
20	10000	1400	120
8	4000	560	48

$$\begin{array}{r} 10000 \\ 1400 \\ 120 \\ 4000 \\ 560 \\ +\quad 48 \\ \hline 16128 \end{array}$$

So, the total cost is £161.28.

Top Tip!

Estimate the answer as a check on your working:
576 x 28p
Round this to 600 x 30p
= 18 000p
= £180

Top Tip!

Check an answer to a division by multiplying by the original divisor:

18 x 32 =
10 x 30 = 300
10 x 2 = 20
8 x 30 = 240
8 x 2 = 16
576

b *Using the standard method:*

$$\begin{array}{r} 32 \\ 18\overline{)576} \\ -\ 540 \\ \hline 36 \\ -\ 36 \\ \hline 0 \end{array}$$

So, they have 32 trays.

Did You Know?

The oldest surviving calculating aid is the Salamis tablet which was used in 300 BC. It is now in the National Museum in Athens.

Spot Check

1 Work out 37 x 52

2 Work out 918 ÷ 17

NUMBER Negative numbers

level 5

Adding and subtracting negative numbers

- To add and subtract negative numbers, use a **number line**.
- Starting at zero, count to the **left** for **negative** numbers and to the **right** for **positive** numbers.

Example: Work out –5 + –6

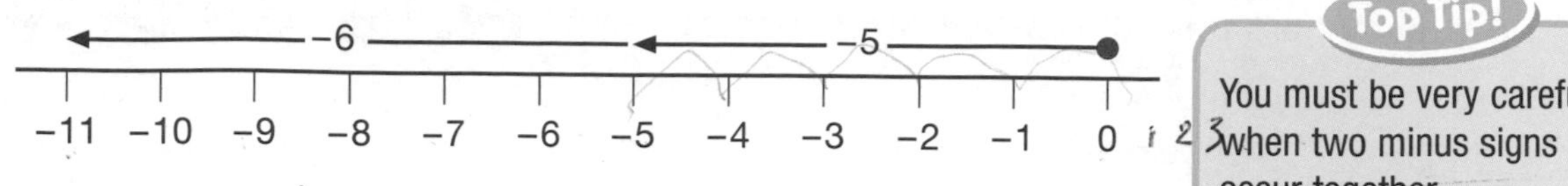

So, –5 + –6 = –11

Top Tip!

You must be very careful when two minus signs occur together.
Two minus signs together act as a **plus**. So +8 – –5 is the same problem as +8 + 5 = 13.

Example: Work out –8 – –9
This is the same as –8 + 9.

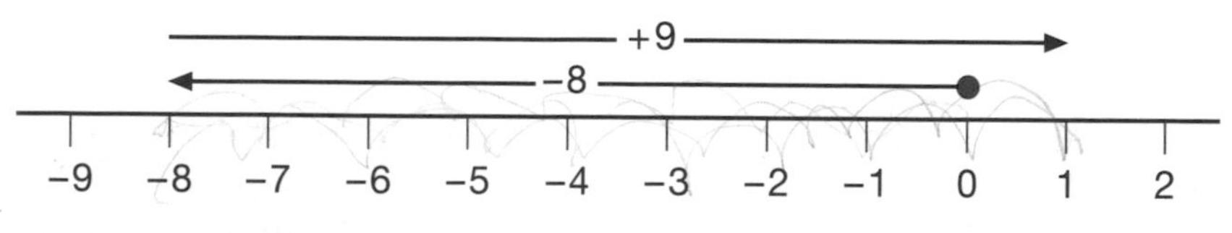

So, –8 – –9 = +1

Multiplying and dividing negative numbers

- Multiplying and dividing negative numbers is just like normal multiplying and dividing. All you have to do is combine the signs together correctly.
- The rules for the signs are the same as for adding and subtracting:

  \+ and – together give a – answer

  – and – together give a + answer.

Example: Work out **a** –3 x +4 **b** –12 ÷ –2

a –3 x +4 is the same as –+3 x 4 = –12
b –12 ÷ –2 is the same as – –12 ÷ 2 = +6

Top Tip!

When multiplying or dividing, if the signs are **different** the answer will be **negative**. If the signs are the **same** the answer will be **positive**.

Example: p and q represent whole numbers.

a Find two numbers p and q such that $p \times q = -12$.
b Find two numbers p and q such that $p + q = -12$.
c Find two numbers p and q such that $p \times q = -12$ and $p + q = -1$.

a Any values that work are acceptable, such as –2 x 6, –12 x 1
b Any values that work are acceptable, such as –2 + –10, +2 – 14
c The only values that work at the same time are –4 and +3.

Spot Check **1** Work out **a** –7 + –2 **b** +3 – –6 **c** –3 x –4 **d** +15 ÷ –3

Sample mental test question

What number is 5 less than –3?

As this is a mental question you will have to picture a number line in your head to get an answer of –8.

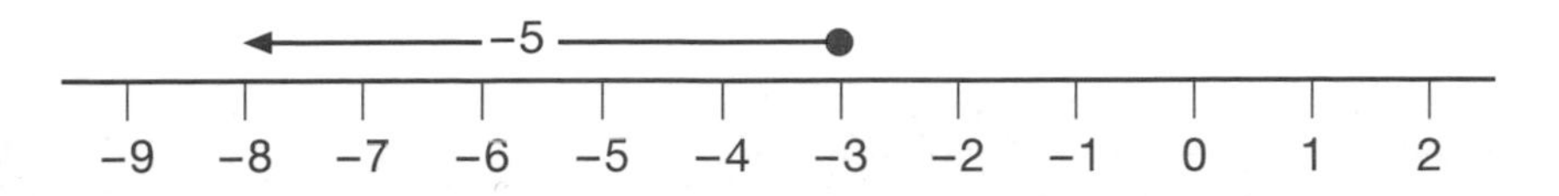

level 6

Sample worked test question

You have these number cards.

a Pick two cards to make the following calculations true.

i .... + ... = 5 **ii** .... – .... = –3 **iii** .... ÷ .... = –2

b Pick two cards to make the answer to the following as large as possible.

 X  = ...

Answers

Try different combinations of the cards.

a ***i*** *6 + –1 = 5* ***ii*** *–1 – 2 = –3* ***iii*** *6 ÷ –3 = –2*

b *–5 x –3 = +15*

Top Tip!

Questions test if you understand the idea of negative numbers. Remember the rules!

Did You Know?

The lowest temperature recorded was –89.4 °C in Vostok, Russia, in 1983 and the highest was 59.4 °C in Libya in 1922.

Adding and subtracting fractions

level 5

Adding fractions

- You can only **add fractions** and **subtract** fractions if they have the same denominator.

Example: Add $\frac{2}{3} + \frac{3}{4}$

First, find the **lowest common multiple** (LCM) of the two denominators, 3 and 4.
The LCM is the smallest common number in the 3 and 4 times table.
This is 12.

Now make both fractions into twelfths.

$\frac{2}{3} = \frac{8}{12}$ (Multiply top and bottom by 4)

$\frac{3}{4} = \frac{9}{12}$ (Multiply top and bottom by 3)

Then just add the numerators and leave the denominator unchanged.

$\frac{2}{3} + \frac{3}{4} = \frac{8}{12} + \frac{9}{12} = \frac{17}{12}$

A fraction like $\frac{17}{12}$ is called **top heavy**. It can be made into a **mixed number** $1\frac{5}{12}$.

level 5

Subtracting fractions

Example: Work out $\frac{5}{9} - \frac{1}{6}$

First, find the **lowest common multiple** (LCM) of the two denominators, 9 and 6. The LCM is the smallest common number in the 9 and 6 times table. This is 18.

Now make both fractions into eighteenths.

$\frac{5}{9} = \frac{10}{18}$ (Multiply top and bottom by 2)

$\frac{1}{6} = \frac{3}{18}$ (Multiply top and bottom by 3)

Then just subtract the numerators and leave the denominator unchanged.

$\frac{5}{9} - \frac{1}{6} = \frac{10}{18} - \frac{3}{18} = \frac{7}{18}$

Top Tip!

Subtracting a fraction from 1 is a common question, e.g.

$1 - \frac{11}{14} = \frac{3}{14}$

$1 - \frac{7}{9} = \frac{2}{9}$

$1 - \frac{8}{11} = \frac{3}{11}$

You should be able to see the relationship between the numerators and denominators.

Spot Check

1 Work out **a** $\frac{3}{5} + \frac{1}{4}$ **b** $\frac{3}{4} - \frac{1}{6}$

2 Work out **a** $1\frac{1}{4} + 2\frac{2}{3}$ **b** $3\frac{3}{5} - 1\frac{3}{8}$

level 6

Mixed numbers

- **Top-heavy** fractions can be converted into **mixed numbers**, which are a combination of a **whole number** and a **fraction**.

Example: Convert **a** $\frac{13}{5}$ to a mixed number **b** $3\frac{2}{9}$ to a top-heavy fraction.

a Divide 13 by 5

$13 \div 5 = 2$ rem 3

So, $\frac{13}{5} = 2\frac{3}{5}$

b Multiply 3 by 9 and add 2

$3 \times 9 + 2 = 29$

So, $3\frac{2}{9} = \frac{29}{9}$

- When adding and subtracting mixed numbers, they can be converted to top-heavy fractions or split into whole numbers and fractions, which makes the calculations easier.

Example: Work out **a** $3\frac{2}{5} + 1\frac{3}{4}$ **b** $4\frac{1}{6} - 2\frac{2}{9}$

a $3\frac{2}{5} + 1\frac{3}{4} = 3 + 1 + \frac{2}{5} + \frac{3}{4} = 4 + \frac{8}{20} + \frac{15}{20} = 4 + \frac{23}{20} = 4 + 1\frac{3}{20} = 5\frac{3}{20}$

b $4\frac{1}{6} - 2\frac{2}{9} = 4 - 2 + \frac{1}{6} - \frac{2}{9} = 2 + \frac{3}{18} - \frac{4}{18} = 2 + -\frac{1}{18} = 1\frac{17}{18}$

Sample mental test questions

Look at the fraction $\frac{4}{12}$. Write it in its simplest form.

The common factor is 4, so divide top and bottom by 4 to get the answer of $\frac{1}{3}$.

Add a half and three quarters.

The fractions will be really easy, so you should know that $\frac{1}{2} + \frac{3}{4} = 1\frac{1}{4}$.

Sample worked test question

A vegetable plot is planted with beans, peas, cabbages and carrots.

The peas take up $\frac{1}{4}$ of the plot.

The beans take up $\frac{3}{8}$ of the plot.

The cabbages take up $\frac{1}{6}$ of the plot.

How much of the plot is planted with carrots?

Answer

The total planted with peas, beans and cabbages is:

$\frac{1}{4} + \frac{3}{8} + \frac{1}{6}$

The common denominator is 24.

Making all the fractions into fractions with a denominator of 24 gives:

$\frac{6}{24} + \frac{9}{24} + \frac{4}{24} = \frac{19}{24}$ *So* $1 - \frac{19}{24} = \frac{5}{24}$ *is planted with carrots.*

Did You Know?

The population of the world increases by $2\frac{2}{3}$ people every second.

Multiplying and dividing fractions

Multiplying fractions

levels 5-6

- Adding and subtracting fractions requires denominators to be the same but multiplying and dividing fractions is a little more straightforward.

Example: Multiply **a** $\frac{2}{3} \times \frac{1}{4}$ **b** $1\frac{1}{4} \times 1\frac{7}{15}$

When multiplying fractions, the new numerator is the product of the numerators and the new denominator is the product of the denominators.

a Multiplying the numerators gives 2 x 1 = 2.
Multiplying the denominators gives 3 x 4 = 12.
So $\frac{2}{3} \times \frac{1}{4} = \frac{2}{12}$.
This fraction $\frac{2}{12}$ will cancel to $\frac{1}{6}$.
To avoid problems with cancelling, cancel any fractions before multiplying:

$$\frac{{}^{1}\cancel{2}}{3} \times \frac{1}{\cancel{4}_{2}} = \frac{1}{6}$$

In this case, 2 on the top cancels with 4 on the bottom by a common factor of 2.

b Write both mixed numbers as **top-heavy** fractions, **cancel common factors** top and bottom and multiply the numerators and denominators. Finally, change the top-heavy answer back into a mixed number.

$$\frac{{}^{1}\cancel{5}}{{}^{2}\cancel{4}} \times \frac{{}^{11}\cancel{22}}{{}^{3}\cancel{15}} = \frac{11}{6} = 1\frac{5}{6}$$

Top Tip!
Always write mixed numbers as top-heavy fractions when multiplying or dividing.
$3\frac{3}{4} = \frac{15}{4}$ because there are 3 x 4 = 12 quarters in 3 plus the extra 3 quarters.

Dividing fractions

level 6

- When dividing fractions, turn the **second** fraction **upside down** and **multiply**.

Example: Divide **a** $\frac{5}{6} \div \frac{1}{3}$ **b** $2\frac{1}{4} \div 1\frac{7}{8}$

a Write $\frac{5}{6} \div \frac{1}{3}$ as $\frac{5}{6} \times \frac{3}{1}$

Once again, cancel if you can. In this case, 3 and 6 cancel by a common factor of 3.

$$\frac{5}{{}^{2}\cancel{6}} \times \frac{{}^{1}\cancel{3}}{1} = \frac{5}{2} = 2\frac{1}{2}$$

Remember to change the top-heavy answer back into a mixed number.

b Write both mixed numbers as **top-heavy** fractions before turning the **second upside down** and **multiplying**. Then cancel common factors top and bottom and multiply the numerators and denominators. Finally, change the top-heavy answer back into a mixed number.

$$\frac{9}{4} \div \frac{15}{8} = \frac{{}^{3}\cancel{9}}{{}^{1}\cancel{4}} \times \frac{{}^{2}\cancel{8}}{{}^{5}\cancel{15}} = \frac{6}{5} = 1\frac{1}{5}$$

Top Tip!
Always cancel before multiplying the numerators and denominators. It makes the calculations much easier.

Sample mental test questions

What is half of one third?

The fractions will be really easy, so you should know that $\frac{1}{2} \times \frac{1}{3} = \frac{1}{6}$.

How many fifths are there in 2?

There are 5 fifths in 1, so there are 10 fifths in 2.

Sample worked test question

level 6

Work out $3\frac{3}{4} \div \frac{5}{8}$

Answers

Write the calculation as $\frac{15}{4} \div \frac{5}{8}$

Now turn the second fraction upside down and multiply $\frac{\cancel{15}^{3}}{\cancel{4}_{1}} \times \frac{\cancel{8}^{2}}{\cancel{5}_{1}} = \frac{6}{1}$

Cancel where possible.

The calculation is $3\frac{3}{4} \div \frac{5}{8} = 6$

Top Tip!

Any fraction with a denominator of 1 is a whole number.

Did You Know?

One mile in every five of the US motorway network has to be straight so that it can be used as an airstrip in emergencies.

Spot Check

1 Work out **a** $\frac{5}{8} \times \frac{4}{15}$ **b** $2\frac{2}{3} \times 1\frac{1}{8}$

2 Work out **a** $\frac{2}{9} \div \frac{8}{15}$ **b** $2\frac{4}{5} \div 2\frac{1}{10}$

Percentages 1

level 5

Percentage parts

- You should know that per cent means 'out of a hundred'.

Example: Work out **a** 15% of £45 **b** 32% of 75 kg

a 10% of £45 is £4.50, so 5% of £45 is £2.25

15% of £45 is £4.50 + £2.25 = £6.75

b On the calculator, do 32 ÷ 100 x 75 = 24 kg or 0.32 x 75 = 24 kg.

level 5

Percentage increases

Example: A car's top speed is 125 mph. After a tune-up, its top speed increases by 12%. What is the new top speed?

Method 1

Work out 12% of 125.

12 ÷ 100 x 125 = 15

Add this to the original speed: 125 + 15 = 140 mph

Method 2

Use a **multiplier**. A 12% increase is a multiplier of 1.12.

1.12 x 125 = 140 mph

Top Tip!

Think of the **per cent** sign as '**÷ 100**' and the '**of**' as a **times** sign. So 43% of 150 is 43 ÷ 100 x 150 = 64.5.

Top Tip!

Calculations are much easier if a 'multiplier' is used.
32% is a multiplier of 0.32.
An increase of 15% is a multiplier of 1.15.
A decrease of 8% is a multiplier of 0.92.
Percentages are easily converted to decimals. Just divide by 100 (or move digits).

level 6

One quantity as a percentage of another

Example: What percentage is 42 out of 56?

The fraction is $\frac{42}{56}$.

This is divided to give the decimal: 42 ÷ 56 = 0.75.

This decimal is multiplied by 100 to give the percentage.

The whole calculation can be done as

42 ÷ 56 x 100 = 75%.

Top Tip!

If you divide the numerator by the denominator this gives the percentage multiplier:
42 ÷ 56 = 0.75, which is 75%.

Example: 17 students in a class of 25 stay for school dinners. What percentage is this?

The fraction is $\frac{17}{25}$.

Multiply the **numerator** (top number) by 4 and multiply the **denominator** (bottom number) by 4.

$\frac{17 \times 4}{25 \times 4} = \frac{68}{100}$

So the answer is 68%.

Sample mental test questions

What is 20% of £30?

Start by calculating 10%, which is £3, and then double it to £6.

In a test I got 16 out of 20. What percentage did I get?

Convert the denominator to 10 or 100.

16 out of 20 is the same as 8 out of 10, which is 80%.

level 5

Sample worked test question

30 students were asked how they travelled to school.

Transport	Boys	Girls
Walk	2	6
Bus	1	9
Car	2	7
Cycle	0	3
Total	5	25

a What percentage of boys come by bus?

b What percentage of girls walk to school?

c Misha said, 'Girls are healthier than boys because more of them walk to school'. Explain why she was wrong.

Answers

a *1 out of 5 is 20%.*

b *6 out of 25 is* $\frac{6}{25} = \frac{24}{100} = 24\%$.

c *Although more girls walk than boys (6 compared to 2), the percentages are* $\frac{2}{5} = 40\%$ *for boys, and* $\frac{6}{25} = 24\%$ *for girls.*

Top Tip!

Make the denominator into 100 by multiplying by a factor of 100 and then do the same thing to the numerator.

Learn the factors of 100:

1 x 100
2 x 50
4 x 25
5 x 20
10 x 10

Did You Know?

In 1971 80% of 8 year olds walked to school. In 2001 the figure was just 7%.

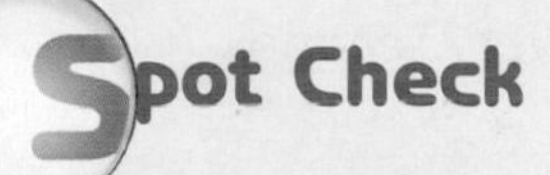

Spot Check

1 Work out **a** £75 increased by 20%

b 120 kg decreased by 12%

NUMBER Percentages 2

level 7

Percentage increases and decreases

- When calculating a percentage increase or decrease, always **divide** by the **original** amount.

Example: Bob's hourly wage goes up from £5.75 to £6.21.
What percentage increase is this?
The actual increase is 6.21 – 5.75 = 0.46.
Percentage increase is 0.46 ÷ 5.75 = 0.08 = 8%.

Example: After suffering from a disease, a seal colony goes from 325 seals to 247.
What percentage decrease is this?
The actual decrease is 325 – 247 = 78.
Percentage decrease is 78 ÷ 325 = 0.24 = 24%.

level 7

Compound interest

- **Compound interest** is calculated using the formula:

$V = P(1 \pm \frac{r}{100})^n$

V is the final amount.

P is the original amount.

± depends on if it is increasing or decreasing.

r is the percentage rate.

n is the number of time periods over which the interest is calculated.

$1 \pm \frac{r}{100}$ is the multiplier you met on page 14.

Example: Alan invests £3000 in an account that pays 4% per annum compound interest. How much does he have after 5 years?
V = £3000 x $(1.04)^5$ = £3649.96

Example: The population of a rural village decreases by 6% per year. In the year 2000, there were 560 inhabitants. How many people live in the village in 2006?
V = 560 x $(0.94)^6$ = 386

Round the final answer off to 2 decimal places for money or a whole number for people.

1 Work out **a** the percentage increase from 80 to 92
b the original cost of a jacket that is now priced at £52.80 after a 12% reduction.

2 To test a detergent, a specific quantity is added to a dish containing 300 000 bacteria.
The detergent kills 20% of the bacteria every second.
How many bacteria are left after 10 seconds?

level 7

Reverse percentages

- **Reverse percentages** are used to find 'the original amount or value'.

Example: After a 12% increase the price of a car is £6160. How much was it before the price increase?

A 12% increase is a multiplier of 1.12.

£6160 is 112% of the original price so the original price is £6160 ÷ 1.12 = £5500.

Example: In a sale, the price of all items is reduced by 15%. A kettle is now priced at £28.90.

What was the original price of the kettle?

A 15% decrease is a multiplier of 0.85.

£28.90 is 85% of the original price, so the original price is £28.90 ÷ 0.85 = £34.

Sample mental test question

After a 10% increase, the price of petrol was 88p per litre. What was the price before the increase?

88p is 110%. It should be obvious that 8p is 10% of 80p, so the original price was 80p per litre.

level 7

Sample worked test question

a Mark invests £2000 in a bank account that pays 5% compound interest for 6 years. Calculate how much he will have after the 6 years.

b After 1 year Jamal had £2310 in the bank. How much did he invest?

Answers

a *A 5% interest rate is a multiplier of 1.05.*
So, after 6 years Mark will have $£2000 \times 1.05^6 = £2680.19$.

b *£2310 represents 105%.*
So, the original amount was 2310 ÷ 1.05 = £2200.

Did You Know?

In the last thousand years, the population of Britain has increased from 2 million to 60 million.

Ratio

level 5

Ratio

- Ratio is a way of **comparing quantities**.
  For example, if there are 18 girls and 12 boys in a class, the ratio is 18 : 12.
  Because 18 and 12 have a **common factor** of 6, this can be cancelled down to 3 : 2.
  This is called the **simplest form**.

Example: Reduce the ratio 15 : 25 to its simplest form.
The highest common factor of 15 and 25 is 5.
Cancelling (dividing) both numbers by 5 gives 3 : 5.

- You may also be asked **direct proportion** questions.

Example: If 6 pencils cost £1.32, how much will 10 pencils cost?

Using **ratio**:
6 : 132 cancels to 1 : 22
Multiplying by 10 gives 10 : 220, so 10 pencils will cost £2.20.

Using the **unitary method**:
If 6 pencils cost £1.32, 1 pencil costs 1.32 ÷ 6 = £0.22.
So 10 pencils cost 10 x £0.22 = £2.20.

These two methods are basically the same.

Top Tip!
Use whichever method is easiest for the question. The unitary method is easier for this example because 6 : 132 is not an easy ratio to cancel.

level 6

Calculating with ratio

- You need to be able to carry out different calculations involving ratios.

Example: If a family of 3 and a family of 2 had a meal and decided to split the bill of £35 between the two families, how much should each family pay?

It wouldn't be fair to split the bill in two, as there are more people in one of the families.
The bill should be split in the ratio 2 : 3.
The ratio 2 : 3 is a total of 2 + 3 = 5 shares.
Each share will be 35 ÷ 5 = 7.
2 x 7 = 14 and 3 x 7 = 21, so the £35 should be split as £14 and £21.

Top Tip!
Always check that the final ratios or values add up to the value you started with, e.g. 14 + 21 = 35.

1 Write the ratio 14 : 18 in its simplest form.
2 Share £32 in the ratio 3 : 5.

Sample mental test questions

Look at the ratio 4 : 10. Write it in its simplest form.
The numbers will make finding a common factor easy.
In this case they cancel by 2 to give 2 : 5.

Divide £100 in the ratio 3 : 7.
The numbers will be easy to divide.
3 + 7 = 10, 100 ÷ 10 = 10, so the shares are £30 and £70.

level
6

Sample worked test question

Aunt Vera decides to give her nephews, Arnie, Barney and Clyde, £180.
The money is to be divided in the ratio of their ages.

Arnie is 1, Barney is 2 and Clyde is 3.

a How much do they each receive?

b The next year she decides to share another £180 between the three boys in the ratio of their ages.
How much do they each receive the following year?

Answers

a *The total of their ages is 6, so divide 180 by 6.*
180 ÷ 6 = 30
Arnie gets 1 x 30 = £30.
Barney gets 2 x 30 = £60.
Clyde gets 3 x 30 = £90.

Top Tip!

Don't forget to check the totals:
30 + 60 + 90 = £180.

b *The following year the ages are 2, 3 and 4. This is a total of 9.*
180 ÷ 9 = 20
Arnie gets 2 x 20 = £40.
Barney gets 3 x 20 = £60.
Clyde gets 4 x 20 = £80.

Did You Know?

The 'Golden ratio' is about 1.618. Leonardo Da Vinci used the proportions of the Golden ratio in the painting of the Mona Lisa.

NUMBER Powers and roots

level 6

Powers

- Powers (or indices) are used to write repeated multiplication problems in a shorter way.
- $3 \times 3 \times 3 \times 3 \times 3 \times 3 = 3^6$. 3 is called the **base number** and 6 is the **power** (or index).
- You can use your calculator to work out powers. It will have a special power key which may look like this. x^y

Example: Which is bigger 4^5 or 5^4?

$4^5 = 4 \times 4 \times 4 \times 4 \times 4 = 1024$

$5^4 = 5 \times 5 \times 5 \times 5 = 625$

So 4^5 is larger.

Top Tip!

When working out a power, always write out the calculation in full and do not get confused. 4^3 is $4 \times 4 \times 4$, not 4×3.

Top Tip!

Square (power 2) and **cube** (power 3) are two special powers which have their own names. You should know the squares up to 15 x 15 and the cubes of 1, 2, 3, 4, 5 and 10.

level 6

Roots

- The **opposite** of a **power** is a **root**. For example $3^4 = 81$ and $\sqrt[4]{81} = 3$. $\sqrt[4]{81}$ is called the fourth root of 81.
- You can use your calculator to work out roots. Square root and cube root usually have their own key: $\sqrt{x}$ and $\sqrt[3]{x}$. Other roots are usually found with: INV and x^y.

Example: Find the following roots **a** $\sqrt[5]{32}$ **b** $\sqrt[3]{125}$

a The 5th root of 32 is 2 because $2 \times 2 \times 2 \times 2 \times 2 = 32$.

b The 3rd (cube) root of 125 is 5 because $5 \times 5 \times 5 = 125$.

Example: Find the values of a and b in these equations.

a $625 = 5^a$

b $\sqrt[b]{729} = 9$

a $5 \times 5 \times 5 \times 5 = 625$, so $a = 4$

b $9 \times 9 \times 9 = 729$, so $b = 3$

Top Tip!

You should know the square roots of all the squares up to 15 x 15 and the cube roots of 1, 8, 27, 64, 125 and 1000.

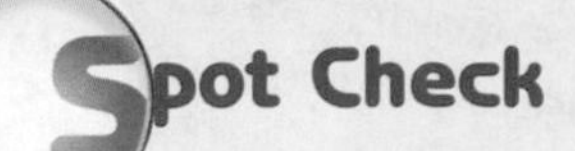

Spot Check

1 Work out the value of **a** 6^3 **b** $\sqrt[4]{16}$

2 Write as a single power of 7 **a** $7^2 \times 7^5$ **b** $7^5 \div 7^2$

level 7

Laws of indices

- When we multiply and divide numbers with indices the following laws apply.

  $a^m \times a^n = a^{m+n}$ **(add the powers)**

  $a^m \div a^n = a^{m-n}$ **(subtract the powers)**

Top Tip!
Any number to the power 0 is 1, e.g. $2^0 = 1$.

Example: **a** Write the following as single powers of 2.

i $2^3 \times 2^5$ **ii** $2^6 \div 2^2$

b Write the following as single powers of x.

i $x^4 \times x^6$ **ii** $x^9 \div x^3$

a i Add the powers: $2^3 \times 2^5 = 2^{3+5} = 2^8$

ii Subtract the powers: $2^6 \div 2^2 = 2^{6-2} = 2^4$

b i Add the powers: $x^4 \times x^6 = x^{4+6} = x^{10}$

ii Subtract the powers: $x^9 \div x^3 = x^{9-3} = x^6$

Sample mental test question

Write down the value of the expression 2^4.

2^4 is $2 \times 2 \times 2 \times 2 = 16$

level 7

Sample worked test question

Look at these cards which show different powers of n.

n^2 | $\sqrt{n}$ | n^3 | $\frac{1}{n^2}$ | $\sqrt[3]{n}$

a When $n = 2$, which cards will give a value lower than 2?

b When $n = 8$, which card gives the greatest value?

c What value of n gives the same value for every card?

Answers

a *Three cards will give a value less than 2.*

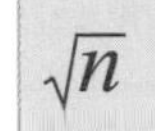

$\frac{1}{n^2}$

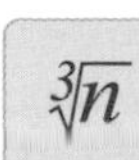

b *n^3 will give the greatest value (512).*

c *$n = 1$ gives the same value (1) for each card.*

Did You Know?
A googol is 10^{100} which is 1 followed by 100 zeros.

Approximations and limits

level 7

Rounding to significant figures

- The number of **significant figures** a number has is simply the number of non-zero digits (zeros between digits count).

  For example, 1700 has 2 significant figures, 3005 has 4 significant figures and 0.08 has 1 significant figure.

  You only need to be able to round numbers to 1, 2 and 3 significant figures.

Example: Round the following numbers to the number of significant figures shown.
a 2672 (2 s.f.) **b** 0.38 (1 s.f.) **c** 1129 (3 s.f.) **d** 4.99 (2 s.f.)
e 5.101 to (3 s.f.)

a 2672 is 2700 to 2 significant figures.

b 0.38 is 0.4 to 1 significant figure.

c 1129 is 1130 to 3 significant figures.

d 4.99 is 5.0 to 2 significant figures.

e 5.101 is 5.10 to 3 significant figures.

- Note how in the last two examples zeros are included after the decimal point to make up the significant figures.

level 7

Approximations

- It is useful to be able to **approximate** the answer to calculations. This way you can check if your answers are correct.

Example: By rounding these numbers to 1 significant figure find an approximate answer to

$$\frac{312 \times 58.2}{19.3}$$

Rounding the numbers to 1 significant figure makes the calculation

$$\frac{300 \times 60}{20}$$

20 goes into 60 three times, so the calculation becomes
300 x 3 = 900.

So, an approximate answer is 900.

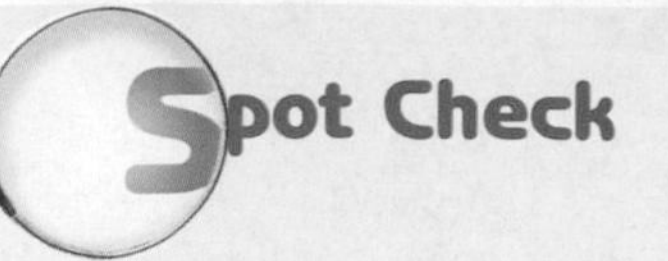

1 Find approximate answers to **a** 178 x 32 **b** 306 ÷ 48

2 The population of a village is given as 370 to the nearest 10.

  a What is the smallest possible population of the village?

  b What is the largest possible population of the village?

levels 7-8

Limits

- When numbers are rounded, the original value could have **between two values** known as the **limits**.

Example: If a plank of wood is measured as 90 cm to the nearest 10 cm, between what limits could the actual length lie?

The lower limit is 85 cm and the upper limit is 95 cm.

Even though 95 cm rounds up to 100 cm, it is still the upper limit.

You write it down as 85 cm ≤ length < 95 cm.

Top Tip!

You can write the upper limit as $94.\dot{9}$.

Example: A practice pitch is measured as 90 metres by 50 metres, each measurement to the nearest metre. John runs round the perimeter of the pitch 10 times. What is the greatest distance he could have run?

The limits of the sides are 89.5 m ≤ length < 90.5 m
49.5 m ≤ width < 50.5 m

The largest perimeter is 90.5 + 50.5 + 90.5 + 50.5 = 282 m.

If John runs round 10 times he could run as far as 10 x 282
= 2820 metres.

Sample mental test question

The mass of a statue is 70 kg to the nearest 10 kilograms.
What is the smallest possible mass it could be?

If the mass is rounded to the nearest 10 then the smallest mass it could be is 65 kg.

level 8

Sample worked test question

A trailer can carry a safe load up to 1000 kg.
Jack has some bags of sand to carry.
Each bag weighs 30 kg to the nearest kilogram.

a What are the limits of the mass of each bag of sand?
b How many bags of sand could the trailer carry safely?

Answers
a *29.5 kg ≤ mass < 30.5 kg*
b *Assume all the bags are the biggest possible mass.*
1000 ÷ 30.5 = 32.78, so 32 is the safe number of bags.

Did You Know?

Each year Florida's beaches lose enough sand to fill Wembley Stadium 15 times.

NUMBER Standard form

level 8

Numbers in standard form

- Standard form is a way of writing very large and very small numbers in a concise way.

  For example, 3 million is 3×10^6 and 0.00005 is 5×10^{-5}.

Example: Write the following standard form numbers as ordinary numbers.
a 3.2×10^4 **b** 4.7×10^{-5}

a $3.2 \times 10^4 = 3.2 \times 10\ 000 = 320\ 000$

b $4.7 \times 10^{-5} = 4.7 \times 0.00001 = 0.000047$

Example: Write the following numbers in standard form.
a 847 000 000 **b** 0.0000005 **c** 24×10^5 **d** 0.52×10^5

a $847\ 000\ 000 = 8.47 \times 100\ 000\ 000 = 8.47 \times 10^8$

b $0.0000005 = 5 \times 0.0000001 = 5 \times 10^{-7}$

c $24 \times 10^5 = 2.4 \times 10 \times 10^5 = 2.4 \times 10^6$

d $0.52 \times 10^5 = 5.2 \times 10^{-1} \times 10^5 = 5.2 \times 10^4$

Top Tip!

Standard form numbers are always $a \times 10^n$ where a is a number between 1 and 10 and n is a positive or negative whole number.

Top Tip!

The power of 10 is the number of times the decimal point moves.

level 8

Calculating with standard form

Example: Work out the following giving your answer in standard form.
a $2.3 \times 10^4 \times 5 \times 10^2$ **b** $4.2 \times 10^5 \div 8 \times 10^3$

a $2.3 \times 10^4 \times 5 \times 10^2 = 2.3 \times 5 \times 10^4 \times 10^2$
$= 11.5 \times 10^6$
$= 1.15 \times 10^7$

b $4.2 \times 10^5 \div 8 \times 10^3 = 4.2 \div 8 \times 10^5 \div 10^3$
$= 0.525 \times 10^2$
$= 5.25 \times 10$

Example: How many seconds are there in a year of 365 days?
Give your answer in standard form to 3 significant figures.

There are 24 hours in a day, 60 minutes in an hour and 60 seconds in a minute.

In 365 days, there are $365 \times 24 \times 60 \times 60$ seconds
$= 31\ 536\ 000$ seconds $= 3.15 \times 10^7$ seconds.

Top Tip!

On some calculators, when the display shows a number like 1.7^{-03} it means 1.7×10^{-3}.

Sample mental test question

Look at the picture of the calculator display.
Circle the number it represents.

8.0-04

8.0004   80 000   0.00008   0.0008

The display is showing 8×10^{-4} so the answer is 0.0008.

level **8**

Sample worked test question

Light takes $8\frac{1}{2}$ minutes to reach the Earth from the Sun.
The speed of light is 186 000 miles a second.
How many miles is the Earth from the Sun?
Give your answer in standard form to 2 significant figures.

Answer
Time taken is $8.5 \times 60 = 510$ seconds.
Distance $= 510 \times 18\ 6000$
$= 94\ 860\ 000$
$= 9.5 \times 10^7$ miles

Did You Know?

The Voyager spacecraft launched in 1977 has passed the edge of our solar system, 8.4×10^9 miles away, and is still sending signals back.

1. Write the following as standard form numbers
   a 0.00007 **b** 7 million
2. Write the following as ordinary numbers
   a 6.2×10^{-6} **b** 5.2×10^9
3. Work out $3.2 \times 10^5 \times 4 \times 10^5$ giving your answer in standard form.

Sequences

level 6

nth term

- The nth term of a sequence can be given by an algebraic expression.

Example: The nth term of a sequence is given by $\frac{1}{2}n^2 + 3$.
Write down the first three terms of the sequence.

The first term is when $n = 1$, so first term $= \frac{1}{2} \times 1^2 + 3 = \frac{1}{2} + 3 = 3\frac{1}{2}$
The second term is when $n = 2$, so second term $= \frac{1}{2} \times 2^2 + 3 = \frac{1}{2} \times 4 + 3 = 5$
The third term is when $n = 3$, so third term $= \frac{1}{2} \times 3^2 + 3 = \frac{1}{2} \times 9 + 3 = 7\frac{1}{2}$

Example: The nth term of a sequence is given by $2n - 1$.
Write down the first five terms of the sequence.

Substitute $n = 1, 2, 3, 4$ and 5 into the rule.
$n = 1$ gives $2 \times 1 - 1 = 1$
$n = 2$ gives $2 \times 2 - 1 = 3$
$n = 3$ gives $2 \times 3 - 1 = 5$
$n = 4$ gives $2 \times 4 - 1 = 7$
$n = 5$ gives $2 \times 5 - 1 = 9$
So the sequence is: 1, 3, 5, 7, 9, ... which are the odd numbers.

level 6

The nth term of a sequence

- The nth term is useful for finding out a number in a sequence without writing out all the sequence.
- There is a quick way of finding out the nth term: work out the difference between consecutive terms and then work out how you get the first number in the sequence from this number.

Example: Find the nth term in the sequence: 4, 9, 14, 19, 24, 29, ...

What does each term go up by? In this case, 5.
The nth term will start $5n$.
What do you do to go from 5 to the first number, 4?
In this case, minus 1.
The nth term will be $5n$ minus 1or $5n - 1$.

Example: Find the nth term of the sequence: 4, 7, 10, 13, 16, ...

The terms increase in steps of 3.
The first term is $3 + 1 = 4$, so the nth term is $3n + 1$.

Example: **a** Write down the nth term of: 1, 4, 9, 16, 25, 36, ...
b Write down the nth term of: 2, 5, 10, 17, 26, 37, ...

a This is the square number sequence so the nth term is n^2.
b This is one more than the square number sequence so the nth term is $n^2 + 1$.

Sample mental test questions

I start at 5 and count down in equal steps: 5, 2, –1. What is the next number in the sequence?

First decide on the step. In this case, it is subtract 3.

Minus 1 subtract 3 is minus 4, so the answer is –4.

The nth term of a sequence is $(n + 2)^2$.

What is the 4th term of the sequence?

Substitute 4 for n:

$(4 + 2)^2 = 6^2 = 36$

Sample worked test question

level 6

The following patterns are made up of black and white hexagons.

Pattern 1

Pattern 2

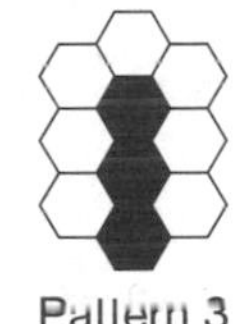

Pattern 3

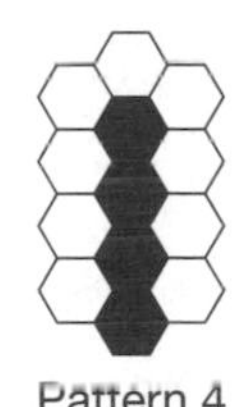

Pattern 4

Complete this table.

Pattern	Black hexagons	White hexagons
5		
10		
n		

Answers

The fifth pattern has 5 black hexagons and 11 white hexagons. You should realise that the number of black hexagons is the same as the pattern number and the number of white hexagons is double the pattern number plus 1.

So the table is:

Pattern	*Black hexagons*	*White hexagons*
5	5	11
10	10	21
n	n	$2n + 1$

Top Tip!

Write out the sequence of numbers as a list. This will help you to see the nth term:

Black hexagons: 1 2 3 4 5

So the nth term is n.

White hexagons: 3 5 7 9 11

So the nth term is $2n + 1$.

Did You Know?

The Fibonacci sequence 1, 1, 2, 3, 5, 8, 13, ... is where each term is formed by the sum of the two preceding numbers. This sequence can be found in nature, for example, in seed heads, flower petals and sea shells.

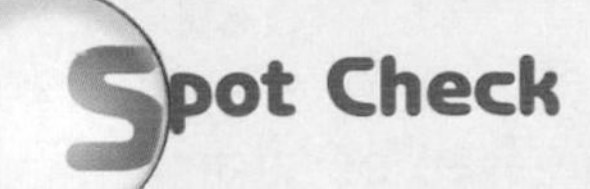

Spot Check

1 What are the next two terms in the sequence: 3, 7, 11, 15, 19, ...

2 What is the nth term of the sequence: 4, 8, 12, 16, 20, ...

3 What is the nth term of the sequence: 5, 9, 13, 17, 21, ...

Square numbers, primes and proof

level 5

Square numbers

Example: Find the next two numbers in this series and describe how the series is built up. 1, 4, 9, 16, 25, ... , ...

The next two terms are 36 and 49.
The series is built up by adding on 3, 5, 7, 9, 11, ...

- Another way of spotting this series is to realise that each number can be written as: 1 x 1, 2 x 2, 3 x 3, 4 x 4, 5 x 5.
  These numbers can be written using a special symbol called **square** or the **power 2** as: 1^2, 2^2, 3^2, 4^2, 5^2.

Top Tip!
The series 1, 4, 9, 16, 25, … is a special series called the **square numbers** because the numbers can be made into square patterns:

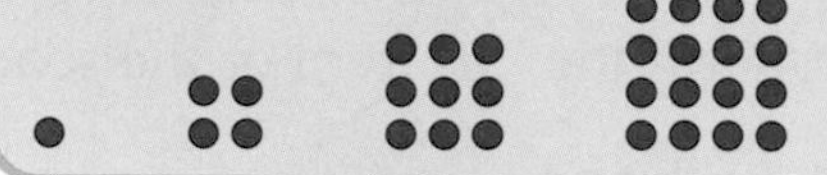

Top Tip!
3^2 is spoken as 'three squared'.

level 5

Prime numbers

- Numbers that only have two factors (1 and itself) are called **prime numbers**.
  There is no pattern to the prime numbers, you just have to learn them or work them out. The prime numbers up to 50 are:
  2, 3, 5, 7, 11, 13, 17, 19, 23, 29, 31, 37, 41, 43, 47

Top Tip!
2 is the only even prime number.

levels 7-8

Proof

- You need to be able to **justify** (or **prove**) results.

Example: p is a prime number, q is an odd number.
State whether the following expressions are: always even, always odd, could be either odd or even.
a pq **b** $2q + 1$ **c** q^2

a p can be even (2) or odd (all other primes), so pq can be either odd or even.
b $2q$ is always even so $2q + 1$ is always odd.
c q^2 is always odd because an odd number multiplied by an odd number is always odd.

Example: n is an integer.
a Explain why $2n + 1$ is always odd.
b Prove that the sum of two odd numbers is always even.

a If n is an integer, either odd or even, then $2n$ must be even because it is multiplied by 2.
If $2n$ is even then $2n + 1$ is odd.
b Let one odd number be $2n + 1$ and the other be $2m + 1$ where n and m are integers.
$2n + 1 + 2m + 1 = 2n + 2m + 2 = 2(n + m + 1)$ which must be even because it has a factor of 2.

Top Tip!
Combinations of odd and even numbers are often tested. To test a rule, pick numbers and try it out.

Sample mental test question

$x^2 = 36$. What are the possible values of $x - 2$?

First, x must be $+6$ or -6, so $+6 - 2 = 4$ and $-6 - 2 = -8$

So there are two answers: -8 and 4.

levels 5-6

Sample worked test question

a Write down the factors of 60.

b Solve the equation $x^2 - 4 = 60$.

c From the list below write down

i a square number **ii** all the prime numbers.

13 15 17 19 21 23

25 27 29 31 33 35

d Write down a number that is

i a prime number that is 1 more than a square number

ii even and a prime number.

Top Tip!

Factors come in pairs except for square numbers, where one number is its own 'pair': 2 x 2, 3 x 3, 1 x 1.

Answers

a *The factors of 60 are 1, 2, 3, 4, 5, 6, 10, 12, 15, 20, 30 and 60.*

b *x^2 must equal 64, so $x = 8$ or -8.*

c ***i*** *The only square number in the list is 25.*

ii *There are several prime numbers in the list: 13, 17, 19, 23, 29 or 31.*

d ***i*** *There many choices e.g. 17 or 37.*

ii *2*

Did You Know?

Each square centimetre of your skin has about 100 000 bacteria on it.

Spot Check

1 p is an odd number. q is an even number. Which of the following are always odd, always even or could be either odd or even?

a p^2 **b** q^2 **c** $p^2 + q^2$ **d** $\frac{2p + q}{2}$

Algebraic manipulation 1

level 5

Expanding brackets

- **Algebra** uses **letters** to **represent values** in equations, expressions and identities. You need to be able to simplify and manipulate algebraic expressions.

Example: Simplify **a** $4a \times 5b$ **b** $2a \times 3a$ **c** $3(2x - 1)$ **d** $4(3x + 2)$
e $3(a + 5b) + 4(a - 3b)$ **f** $6(2a + 1) - 2(3a - 3)$

a $4a \times 5b = 20ab$
b $2a \times 3a = 6a^2$
c $3(2x - 1) = 6x - 3$
d $4(3x + 2) = 12x + 8$
e $3(a + 5b) + 4(a - 3b) = 3a + 15b + 4a - 12b = 7a + 3b$
f $6(2a + 1) - 2(3a - 3) = 12a + 6 - 6a + 6 = 6a + 12$

Top Tip!
When you multiply out a bracket, multiply each term inside the bracket with what is outside the bracket.

level 5

Substitution

- You need to be able to substitute numbers into expressions to find a value.

Example: If $a = -3$, $b = 4$ and $c = -7$, find the value of
a $a + b$ **b** $2a$ **c** $4c - 5$ **d** $a^2 + b^2$ **e** $a(b + c)$

a $a + b = -3 + 4 = 1$
b $2a = 2 \times -3 = -6$
c $4c - 5 = 4 \times -7 - 5 = -28 - 5 = -33$
d $a^2 + b^2 = -3^2 + 4^2 = -3 \times -3 + 4 \times 4 = 25$
e $a(b + c) = -3(4 + -7) = -3 \times -3 = 9$

Top Tip!
Replace the letters with the numbers before doing the calculation.
Don't try to do it in your head.

level 5

Interpreting expressions

- You need to be able to interpret expressions.

Example: Imran is x years old. His sister Aisha is 3 years older. His brother Mushtaq is twice as old as Imran.
a How old is Aisha?
b How old is Mushtaq?
c What is the total of their three ages?
d If the total of their ages is 35 years, how old is Imran?

a Aisha is 3 years older than Imran so she is $x + 3$ years old.
b Mushtaq is twice as old as Imran so he is $2x$ years old.
c The total is $x + x + 3 + 2x = 4x + 3$ years.
d $4x + 3 = 35$, so $4x = 32$, so $x = 8$.
So, Imran is 8 years old.

Top Tip!
A single letter on its own has a coefficient of 1 but there is no need to write it:
$1m = m$

Sample mental test question

Look at the expression $2(a + 5)$. What is the value of the expression when $a = -2$?

Work out the bracket first $a + 5 = -2 + 5 = 3$, then multiply by 2.
The answer is $2 \times 3 = 6$.

level 5

Sample worked test question

There are n students in Form 9A.

a These expressions show how many students are in Forms 9A, 9B and 9C.

9A	n	students
9B	$n + 1$	students
9C	$n - 4$	students

Write the number of students in 9C in words.
How many students are in the three forms altogether?
Give your answer in terms of n.

b Altogether there are 87 students in the three forms.
What is the value of n?

Answers

a *There are four fewer students in Form 9C than Form 9A.*
Total number of students in the three forms = $n + n + 1 + n - 4 = 3n - 3$

b $3n - 3 = 87$
$3n = 90$
$n = 30$

Did You Know?

On August 21, 1965, Charlton Athletic's Keith Peacock became the first substitute to appear in the Football League.

Spot Check

1 Simplify **a** $4(a + 3) - 2(3a + 1)$ **b** $4b \times 5b$
2 If $x = -3$ and $y = 4$, work out **a** $4x + 5y$ **b** $x^2 + y^2$

Algebraic manipulation 2

level 6

Powers

- When manipulating numbers, letters and powers, deal with the **numbers** and each **letter separately**.

Example: Simplify **a** $3a^2b \times 2a^3b^2$ **b** $\dfrac{6ab^2 \times 3a^3b^2}{4a^2b}$

a Treat the numbers, powers of a and powers of b separately.
$3 \times 2 \times a^2 \times a^3 \times b \times b^2 = 6a^5b^3$

b Working out the top row gives $18a^4b^4$.
$\frac{18}{4}$ cancels to $\frac{9}{2}$, $a^4 \div a^2 = a^2$, $b^4 \div b = b^3$

So, the final answer is $\dfrac{9a^2b^3}{2}$.

Top Tip!

Don't forget that only the powers add when multiplying, 3 x 2 is still 6!

level 7

Expanding two linear brackets

- You need to be able to expand a pair of linear brackets.
- There are **three methods** which are shown below.

Method 1: Linear expansion

Example: Expand $(x - 2)(x + 3)$

$$(x - 2)(x + 3) = x(x + 3) - 2(x + 3)$$
$$= x^2 + 3x - 2x - 6$$
$$= x^2 + x - 6$$

Method 2: Box method

Example: Expand $(x + 1)(x - 4)$

	x	$+1$
x	x^2	$+1x$
-4	$-4x$	-4

Collecting up the terms from the boxes:

$$(x + 1)(x - 4) = x^2 + x - 4x - 4$$
$$= x^2 - 3x - 4$$

Method 3: FOIL

- This stands for **F**irst, **O**uter, **I**nner, **L**ast and is the order of multiplying terms.

Example: Expand $(x - 2)(x - 5)$

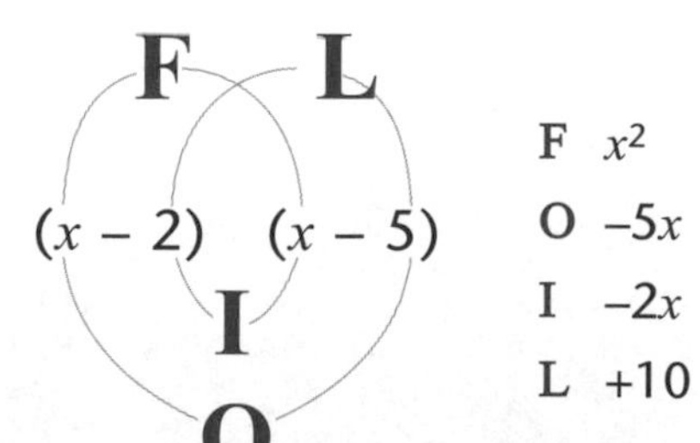

F x^2
O $-5x$
I $-2x$
L $+10$

Collecting the list of terms together:

$(x - 2)(x - 5) = x^2 - 5x - 2x + 10 = x^2 - 7x + 10$

Top Tip!

Make sure you are happy with whichever method you use.
Note that each method generates four terms, then the two x terms combine together to give a final answer with three terms.

Sample mental test question

Look at the division, $x^6 \div x^2$. Write it as a single power of x.

When dividing powers we subtract, so $x^6 \div x^2 = x^4$.

level 7

Sample worked test question

The diagram shows a rectangle that is $n + 3$ units by $n + 1$ units.
The area of one section is filled in.

	n	$+3$
n	n^2	...
$+1$	...	...

a Complete the areas of the other three sections.

b Write down an expression for the total area of the rectangle.

Answers

a *The three sections are filled in on the diagram below.*

	n	$+3$
n	n^2	$3n$
$+1$	n	$+3$

b *The total area is $n^2 + n + 3n + 3 = n^2 + 4n + 3$*

Did You Know?

When it was travelling at twice the speed of sound, Concorde expanded by 300 mm in length due to heat caused by friction.

Spot Check

1 Simplify **a** $3a^2b^3 \times 2a^3b^2$ **b** $\frac{4ab^2 \times 3a^3b^2}{2ab^2}$

2 Expand **a** $(x + 2)(x - 3)$ **b** $(x - 4)(x - 2)$

Factorisation

One common factor

level 7

- When there is a **single factor** that is common to all the terms in an expression, take it out of the bracket.
- If a whole term in the expression is the common factor, leave 1 inside the bracket.

Example: Factorise **a** $3x + 9$ **b** $x^2 - 4x$

a The common factor of $3x$ and 9 is 3, so $3x + 9 = 3 \times x + 3 \times 3$
$= 3(x + 3)$

b The common factor of x^2 and $-4x$ is x, so $x^2 - 4x = x \times x - 4 \times x$
$= x(x - 4)$

Top Tip!

You can check a factorisation by multiplying out the bracket:
$3(x + 3) = 3x + 9$

Common factors

levels 7-8

- When an expression has numbers and letters, factorise the numbers and each letter separately.
- Once again, if a whole term is the common factor, leave 1 inside the bracket.

Example:

a Factorise $6x^2 + 2x$

a The common factor of 6 and 2 is 2, and the common factor of x^2 and x is x.
So, $6 \times 2 + 2x = 2 \times 3 \times x \times x + 2 \times x$
$= 2x(3x + 1)$

b Factorise $4x^2y^3 + 8xy^2$

b The common factor of 4 and 8 is 4; the common factor of x^2 and x is x; and the common factor of y^3 and y^2 is y^2.
So, $4x^2y^3 + 8xy^2 = 4xy^2 \times xy + 4xy^2 \times 2$
$= 4xy^2(xy + 2)$

Factorising quadratics

level 8

- You need to be able to **factorise** a quadratic expression into a **pair of linear brackets**.

Example: Factorise $x^2 + 3x + 2$
The brackets will be of the form $(x \pm a)(x \pm b)$.
If we expand these brackets we get $x^2 + (a + b)x + ab$.
This means that $a + b = 3$ and $ab = 2$.
$a = 2$ and $b = 1$ are true for these equations.
So $x^2 + 3x + 2 = (x + 1)(x + 2)$.

Example: **a** $x^2 + 2x - 8$ **b** $x^2 - 5x + 4$

a Find two numbers p and q so that $p + q = 2$ and $pq = -8$.
The only values that work are $p = 4$ and $q = -2$.
So $x^2 + 2x - 8 = (x + 4)(x - 2)$.

b Find two numbers p and q so that $p + q = -5$ and $pq = 4$.
The only values that work are $p = -1$ and $q = -4$.
So $x^2 - 5x + 4 = (x - 1)(x - 4)$.

Top Tip!

When you are factorising a quadratic such as $x^2 + ax + b$ into the form $(x + p)(x + q)$ then look for two numbers that add up to a and multiply together to give b. i.e. $p + q = a$ and $pq = b$.

Top Tip!

To check your answer, multiply out the brackets.

level 7

Sample worked test questions

a Which of the expressions below is equivalent to the expression $4a^2b^3 + 12a^3b$?

$4ab(ab^2 + 8a^2)$

$4ab(ab^2 + 3a^2)$

$2ab(2ab^2 + 6a^2)$

$3ab(ab^2 + 4a^2)$

b Which of the expressions below is equivalent to the expression $x^2 + 8x + 12$?

$(x + 3)(x + 4)$

$(x - 3)(x - 4)$

$(x + 2)(x + 6)$

$(x - 2)(x - 6)$

Answers

a *There are two expressions that are equivalent.*

$2ab(2ab^2 + 6a^2)$ *and* $4ab(ab^2 + 3a^2)$

b *Expanding each expression shows that* $(x + 2)(x + 6)$ *is the correct factorisation.*

Did You Know?

The first known quadratic expressions have been found on an Egyptian papyrus dated about 2000 BC.

Spot Check

1 Factorise **a** $3x^2 + 6x$ **b** $4a^2b^3 + 6a^3b$

2 Factorise **a** $x^2 - 9x + 20$ **b** $x^2 + 3x - 10$

ALGEBRA Formulae

Formulae

level 5

- A **formula** is a **rule** that changes one number into another.
- A **flow diagram** can be a useful way to represent a **formula**.

Example: Look at the flow diagram below.

a If the input is 6 what is the output?

b What is the input if the output is 21?

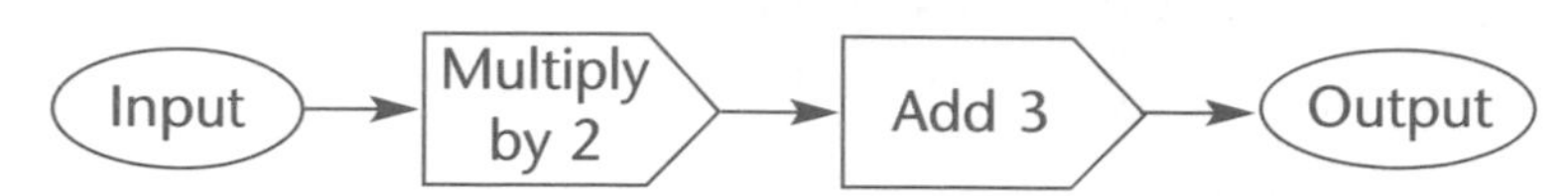

a An input of 6 gives an output of 2 x 6 + 3 = 15.

b To find an input from an output **work backwards** through the flow diagram.

An output of 21 has an input of (21 – 3) ÷ 2 = 9.

Top Tip!

A flow diagram can be used to help solve equations:

$2x + 3 = 21$

$x = 9$

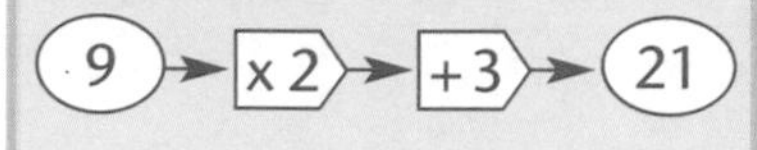

Example: To cook a turkey, use this formula:

Cooking time (in min) = Weight of turkey (in kg) x 30 min + 30 min

a How long will it take to cook a 6 kilogram turkey?

b If a turkey took 5 hours to cook, how much did it weigh?

a Cooking time = 6 x 30 min plus 30 min
= 210 minutes
= 3 hours and 30 min

b 5 hours = 300 minutes.

Deduct 30 minutes, 300 – 30 = 270 minutes

Divide by 30, 270 ÷ 30 = 9

The turkey weighed 9 kilograms.

Top Tip!

Try to write the rule in symbols:

$T = W \times 30 + 30$

So, if $W = 6$,

$T = 6 \times 30 + 30 = 210$

level 8

Example: Find the values of a and b when $p = 10$.

i $a = \frac{3p^2}{5}$ **ii** $b = \frac{2p^3(p-4)}{25}$

i $a = \frac{3 \times 10^2}{5} = \frac{3 \times 100}{5} = \frac{300}{5} = 60$

ii $b = \frac{2 \times 10^3 \times (10-4)}{25} = \frac{2 \times 1000 \times 6}{25} = 2 \times 40 \times 6 = 480$

Top Tip!

Substitute numbers into the formula before trying to work it out.

1 I think of a number, multiply it by 3 and subtract 4. The result is 8 more than the number I first thought of. What was the number I thought of?

2 Work out the value of $\sqrt{a^2 + b^2}$ when $a = 3$ and $b = 4$.

Sample mental test question

Look at this flow diagram. What is the input if the output is 7?

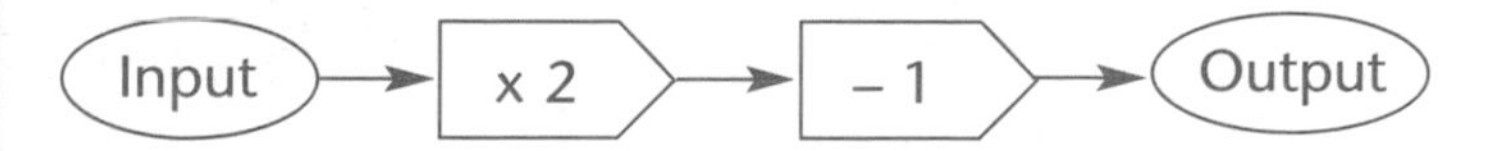

*You should know to work **backwards** through the flow diagram.*

$7 + 1 = 8$, $8 \div 2 = 4$

So the input is 4.

Sample worked test question

level 8

The perimeter (P) and area (A) of this shape are given by

$P = 3a + \sqrt{a^2 + 2b^2 - 2ab}$

$A = a^2 - \frac{ab}{2} + \frac{b^2}{2}$

a
b
a
a – b

a Work out the perimeter and area when $a = 5$ cm and $b = 7$ cm.

b Work out the perimeter and area when $a = 6$ cm and $b = 12$ cm.

Answers

a $P = 3 \times 5 + \sqrt{(5^2 + 2 \times 7^2 - 2 \times 5 \times 7)}$

$= 15 + 7.28 = 22.3$ cm

$A = 5^2 - \frac{5 \times 7}{2} + \frac{7^2}{2} = 25 - 17.5 + 24.5 = 32$ cm^2

b $P = 3 \times 6 + \sqrt{(6^2 + 2 \times 12^2 - 2 \times 6 \times 12)}$

$= 18 + 13.42 = 31.4$ cm

$A = 6^2 - \frac{6 \times 12}{2} + \frac{12^2}{2} = 36 - 36 + 72 = 72$ cm^2

Did You Know?

Toothpaste was first sold commercially in 1873 but the earliest formula for toothpaste was written in the fourth century. (It contained soot!)

ALGEBRA Graphs 1

x and y lines

levels 5-6

- There are some graphs that you need to learn.

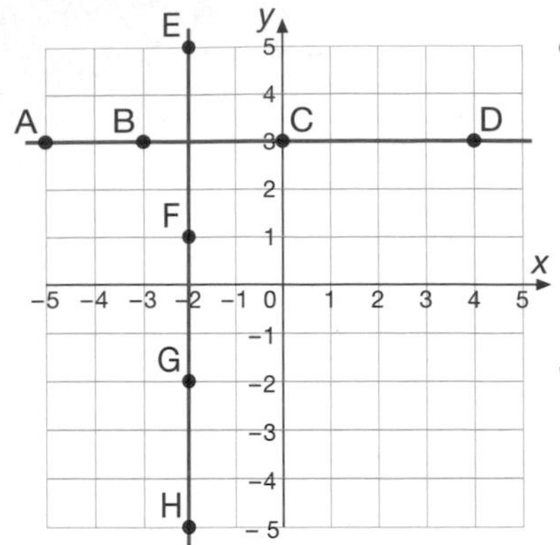

- The coordinates of the points A, B, C and D are (–5, 3), (–3, 3), (0, 3), (4, 3) respectively.
  You can see that they all have a y-coordinate of 3 and form a straight line on the grid.
  This line has an equation $y = 3$.

- The coordinates of the points E, F, G and H are (–2, 5), (–2, 1), (–2, –2), (–2, –5) respectively.
  You can see that they all have an x-coordinate of –2 and form a straight line on the grid.
  This line has an equation $x = -2$.

Top Tip!

All lines of the form $y = b$ are **horizontal**, and lines of the form $x = a$ are **vertical**.

The x-axis is the line $y = 0$
The y-axis is the line $x = 0$

x and y graphs

level 5

- Two other graphs that you need to know are:

$y = x$

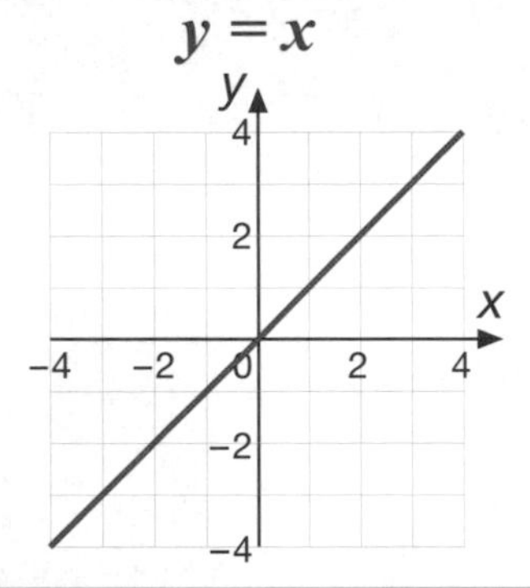

$y = -x$

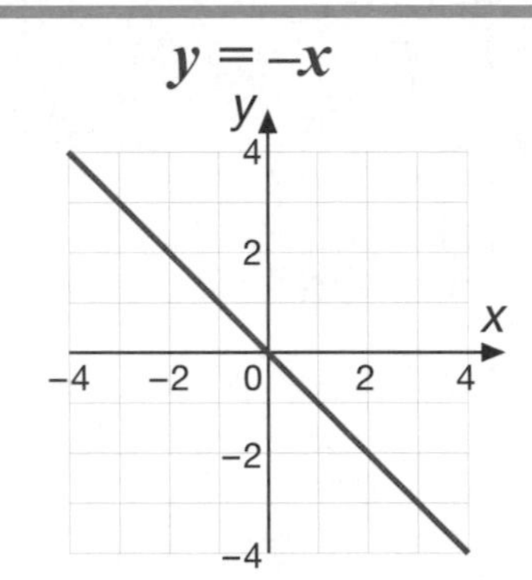

Drawing graphs by plotting points

levels 5-6

- Graphs show the relationship between variables on a coordinate grid.
  For example, the equation $y = 2x + 1$ shows a relationship between x and y where the y-value is 2 times the x-value plus 1.
  If $x = 0$, $y = 2 \times 0 + 1 = 1$. This can be represented by the coordinates (0, 1).
- Similarly, when $x = 1$, $y = 2 \times 1 + 1 = 3$. This is the point (1, 3).
  Other coordinates connecting x and y are (–3, –5), (–1, –1), (2, 5).
  When these are plotted on a graph, they can be joined by a straight line.
- Coordinates are always given in the order: (x, y).

Top Tip!

Always label graphs.

Example: Draw the graph of $y = 3x - 1$.

First find some points by choosing x-values:

Let $x = 0$, $y = 3 \times 0 - 1 = -1$

Let $x = 1$, $y = 3 \times 1 - 1 = 2$

Let $x = 2$, $y = 3 \times 2 - 1 = 5$

Let $x = -1$, $y = 3 \times -1 - 1 = -4$

Plot the points and join them up.

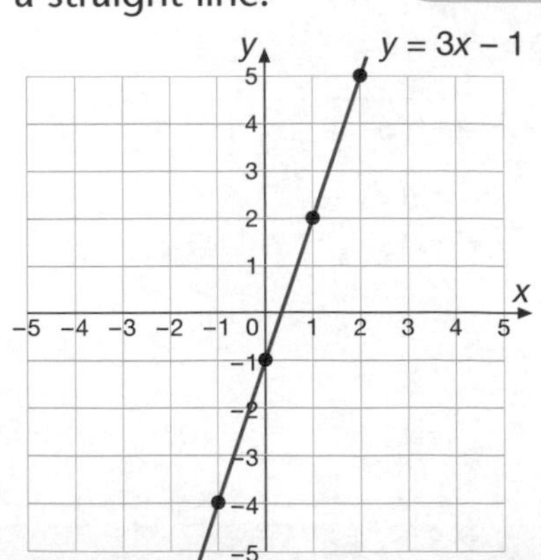

level 6

$x + y = c$ graphs

- The coordinates of the points A, B, C and D are (–3, 5), (0, 2), (3, –1), (5, –3) respectively.

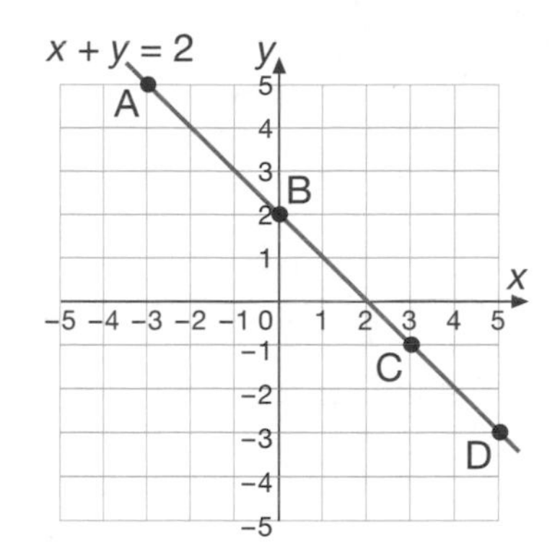

You can see that the x- and y-coordinates add up to a total of 2. This line has an equation $x + y = 2$.

Top Tip!

All lines of the form $x + y = c$ slope at 45° from top left to bottom right, and pass through the value c on both axes.

Sample mental test question

Look at the equation $y = 3x + 2$. What is the value of y when $x = 2$?

Substitute $x = 2$ into the equation, so $y = 3 \times 2 + 2 = 8$.

level 6

Sample worked test question

A is the point (2, 4).
B is the point (–4, –2).

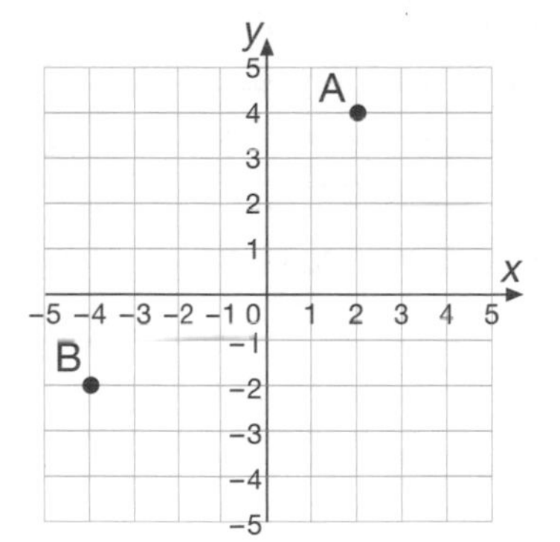

Which of the following equations is the graph of the straight line through A and B?

$y = 2x$ $\quad y = x + 2$ $\quad y = -2x$ $\quad y = \frac{x}{2}$

Explain your answer.

Answer

The equation is $y = x + 2$. This is the only equation that fits both A and B.

For A: $4 = 2 + 2$ For B: $-2 = -4 + 2$

Top Tip!

Substitute the x- and y-values into the equations for all the pairs of coordinates you are given.

Did You Know?

René Descartes devised x- and y-coordinates in the seventeenth century. The grid is also called the Cartesian plane after Descartes.

Spot Check

1 Complete the table for $y = 3x - 2$ for values of x from –2 to +4.

x	–2	–1	0	1	2	3	4
y	–8						10

ALGEBRA Graphs 2

Drawing graphs by the gradient–intercept method

level 6

- In an equation like $y = mx + c$, c is where the graph crosses the y-axis and m is the gradient.

  So for $y = 3x - 1$, the graph crosses the y-axis at –1 and has a gradient of 3. This means for every 1 unit across, the graph goes up by 3 units.

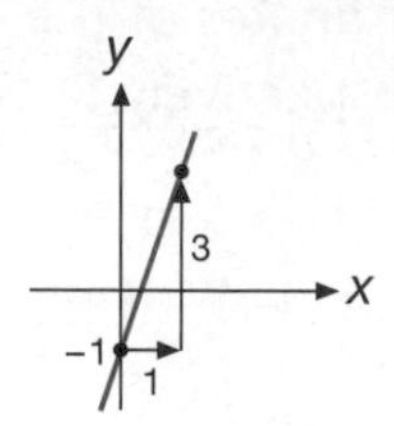

Example: Draw the graph of $y = 2x - 1$.

Start by plotting the point (0, –1).

Then from (0, –1) count 1 square across and 2 squares up, mark a point, repeat from this point and so on. You can also count 1 square back and 2 squares down.

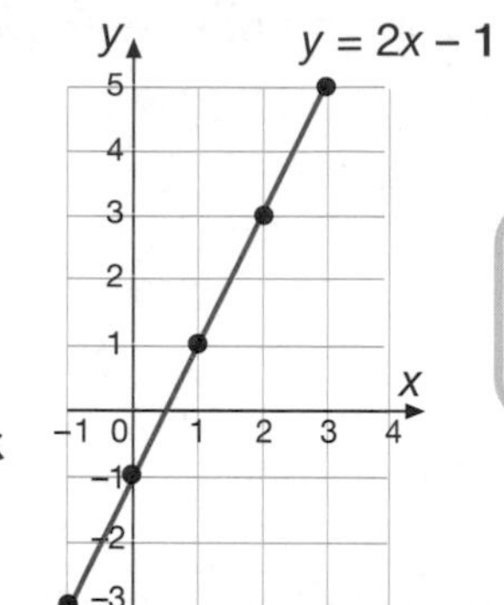

Top Tip!

Remember to read the x-axis first then the y-axis.

Finding the equation of graphs by the gradient–intercept method

level 7

- The **gradient** of a line is given by the ***y*-step divided by** the ***x*-step** between any two points on the line.

Example: Work out the equation of the line passing through (0, 0) and (2, 8).

Between (0, 0) and (2, 8) the y-step is 8 and the x-step is 2.

So, the gradient is $8 \div 2 = 4$.

The intercept of the y-axis is at 0, so the equation of the line is $y = 4x + 0$ or just $y = 4x$.

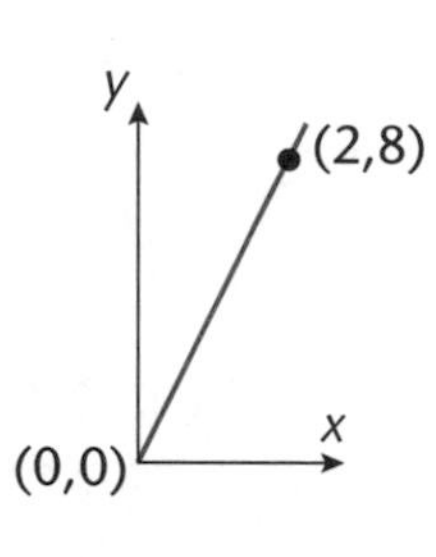

Top Tip!

Lines that slope from top left to bottom right have a **negative gradient**.

Top Tip!

It doesn't matter which two points on the line you choose because the gradient will be the same. However, choose points with coordinates that make your calculations easy.

Non-linear graphs

levels 7-8

- Graphs can be used to show **non-linear** relationships such as **quadratic** graphs.

Example: The table shows values of the equation $y = x^2 + 2x - 1$ for values of x from –3 to +3.

a Complete the table and plot the graph.

x	–3	–2	–1	0	1	2	3
y	2			–1	2		14

b Find the values of x when $y = 1$.

a The missing values from the table are –1, –2 and 7.

These points are plotted and joined to make a smooth curve.

b Reading from the graph, when $y = 1$, $x = -2.7$ and 0.7.

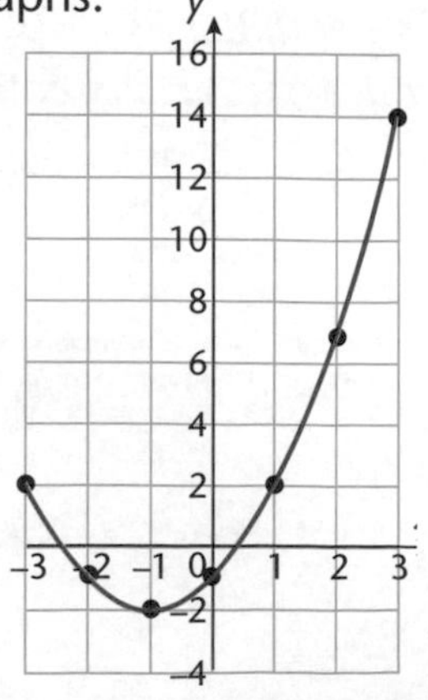

levels 7-8

Real-life graphs

- Graphs can also be used to show **real-life situations** such as distance–time graphs and the depth as containers are filled with water.

Example: A cricket ball is thrown from the boundary back to the wicket keeper. Which of the following graphs shows the path of the ball?

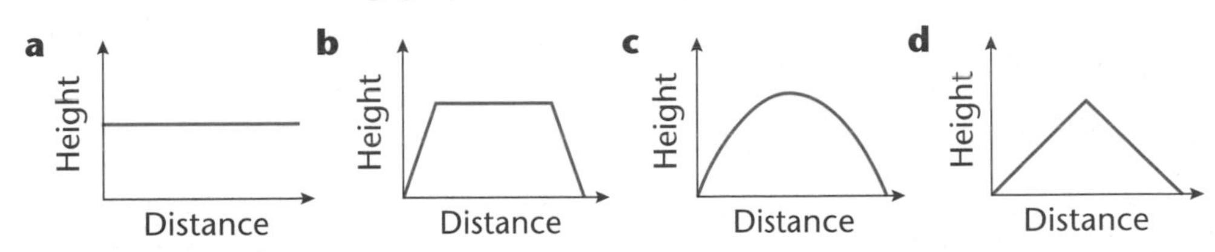

Graph **c** shows the path of the ball. The path of any object thrown in the air is a **parabolic curve**.

Sample mental test question

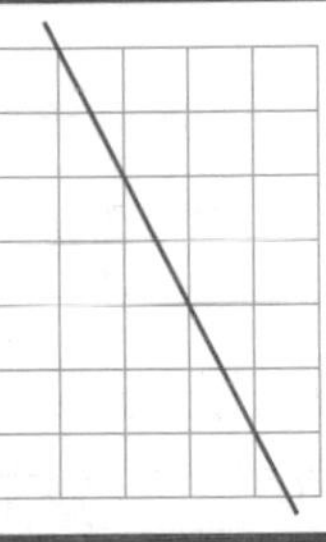

What is the gradient of the line shown?

Starting at any point for every one unit moved across, the line goes ***down*** *by 2 units. This means the gradient is –2.*

level 7

Sample worked test question

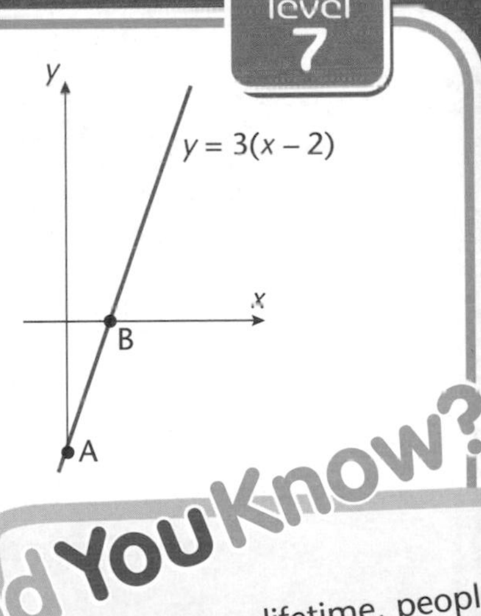

The diagram shows a straight line with equation $y = 3(x - 2)$.
Work out the coordinates of the points A and B.

Answers
The point A is the intercept on the y*-axis.*
Expanding the equation gives $y = 3x - 6$.
So, A is (0, –6).
The point B is the intercept with the x*-axis. The equation of the* x*-axis is* $y = 0$.
When $y = 0$, $0 = 3x - 6$, *so* $x = 2$.
So, B is (2, 0).

Did You Know?
In an average lifetime, people spend six months waiting for red lights to turn green.

Spot Check

Match the following graphs with the equations shown below.

a

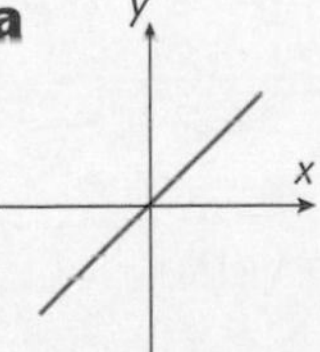

b

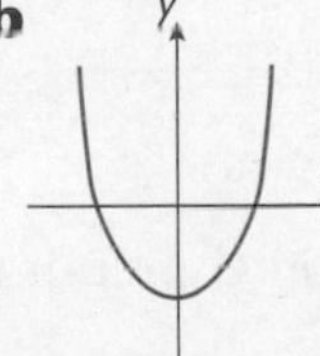

c

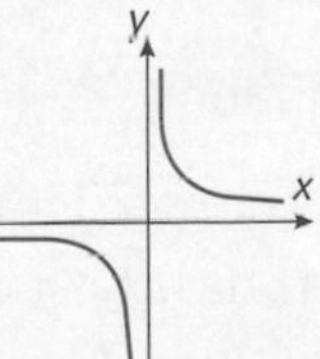

d

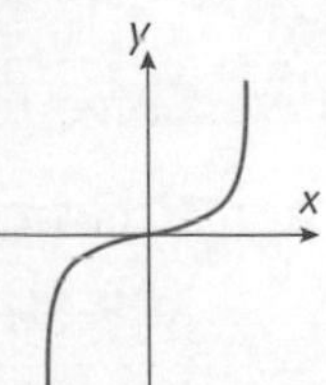

i $y = x^3$ **ii** $y = x^2 - 3$ **iii** $y = x$ **iv** $y = \frac{1}{x}$

ALGEBRA Equations 1

level 5

Basic equations

- 'I am thinking of a number. I double it and add 6. The answer is 12. What was the number I thought of?'
- This type of question can usually be solved in your head to give the answer 3.
  It can also be written as an **equation**.
  An equation is an expression involving a certain letter, x say, that is equal to a number.
  '**Solving the equation**' means **finding** the **value** of x that makes it true.

Example: The puzzle above could be written as: $2x + 6 = 12$

To solve this equation, you need x on its own on the left-hand side of the equals sign. You do this by applying the **inverse operations** to **both sides**.
First, eliminate '+ 6' by doing the inverse operation '– 6':
$2x = 6$
Now eliminate 'x 2' by doing the inverse operation '÷ 2':
$x = 3$

Example: Solve **a** $\frac{x}{2} - 3 = 7$ **b** $\frac{x-3}{2} = 6$

a First add 3, then multiply by 2.

$\frac{x}{2} = 10$
$x = 20$

b First multiply by 2 and then add 3.

$x - 3 = 12$
$x = 15$

Top Tip!

There are different ways of writing the solution to an equation, but they all arrive at the same solution:

$$2x + 6 - 6 = 12 - 6$$
$$2x = 6$$
$$2x \div 2 = 6 \div 2$$
$$x = 3$$

levels 5-6

Fractional equations

- A **fractional equation** is one in which the **variable** appears as the **numerator** or **denominator** of a fraction.

Example: Solve the equation $\frac{x}{2} = \frac{7}{4}$

The first step in solving a fractional equation is to **cross-multiply**. This means multiply the denominator of the left-hand side by the numerator of the right-hand side and multiply the denominator of the right-hand side by the numerator of the left-hand side.
So the equation becomes $4 \times x = 2 \times 7$.
Tidy up the terms: $4x = 14$
Solve the equation: $x = 3\frac{1}{2}$ or 3.5

You could just write down $4x = 14$ and then solve the equation. Answers can be left as top-heavy fractions, such as $\frac{14}{4}$ unless you are asked for an answer in its **simplest form**.

Example: Solve the equation $\frac{2}{x} = \frac{7}{11}$

Cross-multiplying: $22 = 7x$
This equation has the x-term on the right so reverse the equation: $7x = 22$
Divide by 7: $x = \frac{22}{7} = 3\frac{1}{7}$

Top Tip!

You can remember cross-multiplying by thinking of

$$\frac{x}{2} \times \frac{7}{4}$$

which is where the term 'cross' comes from.

Sample mental test question

Look at the equation $3x - 4 = 11$.

What value of x makes the equation true?

If $3x - 4 = 11$, $3x = 15$ and $x = 5$.

level 5

Sample worked test question

Solve the equations to find the values of x, y and z.

$3x + 10$ → $= 40$

$\frac{y}{2} - 10$ → $= 40$

$z^2 + 4$ → $= 40$

Answers

$3x + 10 = 40 \Rightarrow 3x = 30 \Rightarrow x = 10$

$\frac{y}{2} - 10 = 40 \Rightarrow \frac{y}{2} = 50 \Rightarrow y = 100$

$z^2 + 4 = 40 \Rightarrow z^2 = 36 \Rightarrow z = 6$ (or -6)

Did You Know?

The world's largest jigsaw puzzle had 18 240 pieces and took 10 months to solve.

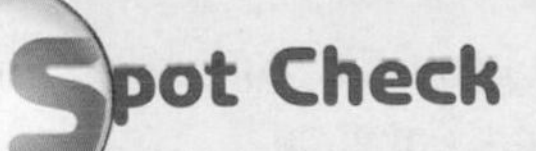

Spot Check

1 Solve the equations **a** $3x - 8 = 10$ **b** $\frac{x}{4} = \frac{5}{2}$

2 Solve the equations **a** $\frac{4x - 1}{3} = 5$ **b** $\frac{3x + 1}{2} = 11$

ALGEBRA Equations 2

level 5

Equations with the variable on both sides

- To solve an equation where the variable is on **both sides**, you need to get all the **variable terms** on the **left-hand side** of the equals sign and the **numbers** on the **right-hand side**.
- You do this by using the rule '**change sides change signs**'.

Example:

I am thinking of a number. I double it and add 5. The answer is 7 more than the number I first thought of. What was the number I thought of?

This type of question can usually be solved in your head, giving the answer 2. It can also be written as an **equation**. The equation for the situation above would be $2x + 5 = x + 7$.

Example: Solve these equations.

a $3x - 7 = x + 5$

b $4x + 3 = 8 - x$

Start by collecting all the x terms on the left-hand side and the numbers on the right-hand side. Remember to use the 'change sides change signs' rule.

a $3x - x = 5 + 7$

This can be simplified to $2x = 12$ so $x = 6$.

b $4x + x = 8 - 3$

This can be simplified to $5x = 5$ so $x = 1$.

Be careful when moving terms across the equals sign. Make sure you remember to change the sign.

level 6

Equations with brackets

- To solve equations with brackets, first **expand** the brackets.
- If the equation has fractions, **cross-multiply** to get rid of these too.

Example: Solve these equations.

a $2(x - 3) = 10$ **b** $\frac{x - 3}{2} = 7$

The second equation may not appear to have a bracket but the line separating the top of the fraction from the bottom acts as a bracket.

a $2x - 6 = 10$

$2x = 16$

$x = 8$

b $x - 3 = 14$

$x = 17$

Top Tip!

Always check your answer in the original equation.

level 7

Inequalities

- Inequalities are solved in much the same way as equations but the answer can be a **range of values**.
- When you show an inequality on a number line, use an **open circle** to show that the boundary is not included and a **filled-in** circle to show that it is.

$<$ means less than

$\leq$ means less than or equal to

Example: Solve these and show their solution on a number line.

a $2x + 3 < 7$ **b** $5(x + 3) \geq 10$

a $2x < 7 - 3$

$2x < 4$

$x < 2$

b $5x + 15 \geq 10$

$5x \geq -5$

$x \geq -1$

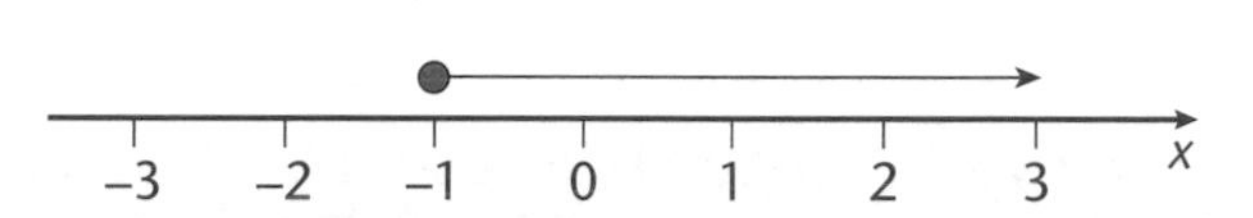

Sample mental test question

Look at the inequality $x > 3$.
What is the smallest integer value of x that obeys the inequality?

If $x > 3$, the smallest whole number that obeys the inequality is $x = 4$.

level 7

Sample worked test question

Solve this equation $\frac{3(2x - 1)}{5x} = 1$

Answer

Multiply out the bracket and cross-multiply the denominator.

$6x - 3 = 5x$

Rearrange to get the x terms on the left and the number on the right.

$6x - 5x = 3$

$x = 3$

Did You Know?

The 350 richest people in the world have more money than the poorest 45% of the world's population (about 3 billion people).

1 Solve these equations **a** $3x - 8 = 10 - x$ **b** $3(x - 5) = 6$

2 Solve these inequalities **a** $4x - 7 > 3$ **b** $\frac{x + 3}{2} \geq 7$

ALGEBRA Trial and improvement

level 6

Trial and improvement

- The only way to solve an equation like $x^3 + 2x = 27$ is by **trial and improvement**.
- Trial and improvement is just sensible **guesswork**.

Example: There is a solution of the equation $x^3 + 2x = 27$ between 2 and 3. Find the solution to 1 decimal place.

Start by making a guess between 2 and 3:

2.5 is a sensible guess. $2.5^3 + 2 \times 2.5 = 20.625$

Then make a better guess: $2.6^3 + 2 \times 2.6 = 22.776$

Keep on making better guesses: $2.7^3 + 2 \times 2.7 = 25.083$

$2.8^3 + 2 \times 2.8 = 27.552$

When you find two 1 decimal place values that 'bracket' the answer, check the middle value to make sure which of the values is closer.

$2.75^3 + 2 \times 2.75 = 26.296\ 875$

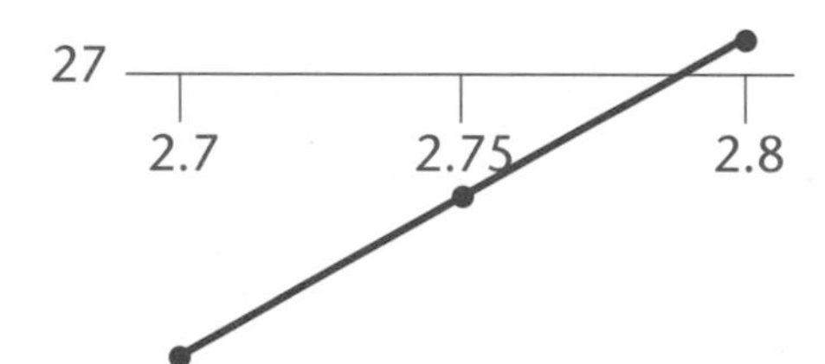

This means that 2.8 is the closer value to the answer.

level 6

Using a table to help you

- Setting out the working in a **table** makes it easier to understand.

Example: Continue the table to solve the equation $x^3 - x = 50$

x	$x^3 - x$	Comment
4	60	Too high

First try the next number below 4, then keep on refining the guess.

x	$x^3 - x$	Comment
4	60	Too high
3	24	Too low
3.5	39.375	Too low
3.8	51.072	Too high
3.7	46.953	Too low
3.75	48.984375	Too low

The nearest 1 decimal place value is $x = 3.8$.

Top Tip!

The best way to set out these problems is in a table. Tables are often given in test questions.

level 6

Sample worked test question

A rectangle has a side of length y centimetres.
The other side is of length $y + 3$ centimetres.

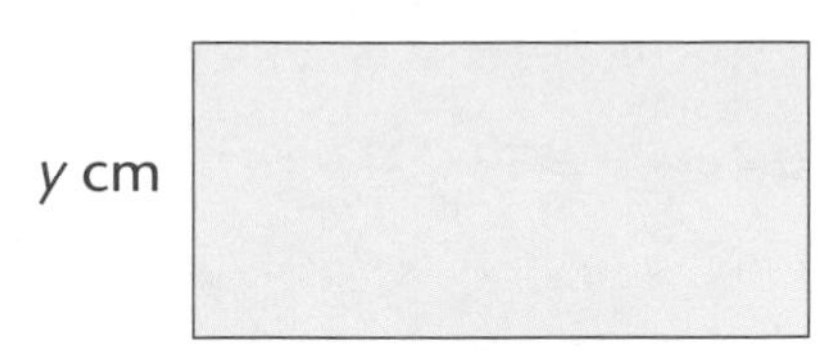

$y + 3$ cm

The area of the rectangle is 48.16 cm^2.
This equation shows the area of the rectangle:
$y(y + 3) = 48.16$
Find the value of y.

y	$y + 3$	$y(y + 3)$	Comment
4	7	28	Too low

Answer
The given starting value of 4 gives an area that is too low.

Continue the table with a higher value than 4.

y	$y + 3$	$y)y + 3)$	Comment
4	*7*	*28*	*Too low*
5	*8*	*40*	*Too low*
6	*9*	*54*	*Too high*
5.5	*8.5*	*46.75*	*Too low*
5.6	*8.6*	*48.16*	*Exact*

Because the answer is exact there is no need to test a halfway value.

Did You Know?
The longest trial in British history was the McLibel trial which lasted three years.

Spot Check

1 Show clearly why there is a solution of the equation $2x^3 + 3x = 100$ between $x = 3$ and $x = 4$.

Simultaneous equations

level 7

Balancing equations

- **Simultaneous equations** are two equations with **two variables** that have a **common solution**.
- The first stage in solving a pair of simultaneous equations is to **balance** the coefficient of one of the variables.

Example: Balance one of the terms of the equations $2x - 3y = 3$ and $4x + y = 13$.

$2x - 3y = 3$ (1)

$4x + y = 13$ (2)

Note that we can balance either the x or y terms.

Equation (1) x 2 gives $4x - 6y = 6$, which balances the x terms.

Equation (2) x 3 gives $12x + 3y = 39$, which balances the y terms.

- Note it doesn't matter if the signs are plus or minus. It is only the **coefficient** that has to be the same.

Top Tip!

Give the equations a 'label' either a letter or a number so you can explain your working more easily.

level 7

Eliminating

- Once one of the terms has been balanced then it can be **eliminated**.
- This is done either by **subtraction** (if the signs are the same) or by **addition** (if the signs are different).

Example: Eliminate one of the terms in these pairs of equations.

a $4x + y = 13$ and $4x - 6y = 6$

b $2x - 3y = 3$ and $12x + 3y = 39$

a This pair of equations has the same x term ($4x$) with the same sign so we will subtract them.

$4x + y = 13$ (1)

$4x - 6y = 6$ (2)

Equation (1) – equation (2) gives $7y = 7$.

So $y = 1$.

b This pair of equations has the same y term ($3y$) with different signs so we will add them.

$2x - 3y = 3$ (1)

$12x + 3y = 39$ (2)

Equation (1) + equation (2) gives $14x = 42$.

So $x = 3$.

level
8

Simultaneous equations

- Simultaneous equations are solved by **balancing terms** and **eliminating** to give a simpler equation which can be solved.

Example: Solve these simultaneous equations.

a $2x + 3y = 9$, $2x - y = 1$

b $5x + 3y = 2$, $2x - 4y = 6$

a $2x + 3y = 9$ (1)

$2x - y = 1$ (2)

These equations already have the same x terms so there is no need to balance them first.

(1) – (2) $4y = 8 \Rightarrow y = 2$

If $y = 2$, using equation (1), $2x + 3 \times 2 = 9 \Rightarrow x = 1.5$

b $5x + 3y = 2$ (1)

$2x - 4y = 6$ (2)

(1) x 2 $10x + 6y = 4$ (3)

(2) x 5 $10x - 20y = 30$ (4)

(3) – (4) $26y = -26 \Rightarrow y = -1$

If $y = -1$, using equation (1), $5x - 3 = 2 \Rightarrow x = 1$

Top Tip!

You can always multiply each equation by the coefficient of x in the other equation to balance the x terms.

level
8

Sample worked test question

Solve the simultaneous equations $4x + 3y = 3$ and $2x - 5y = 8$.

Answers

$4x + 3y = 3$ (1)

$2x - 5y = 8$ (2)

(2) x 2 $4x - 10y = 16$ (3)

(1) – (3) $13y = -13 \Rightarrow y = -1$

If $y = -1$, using equation (1), $4x - 3 = 3 \Rightarrow x = 1.5$

Did You Know?

John Evans of Derbyshire holds thirty world records for balancing things on his head, including a Mini Cooper car!

1 Solve the simultaneous equations $3x - 8y = 14$ and $5x - 2y = 12$.

SHAPE, SPACE AND MEASURES

Bearings

level 6

Bearings

Example: Measure **i** the distance and **ii** the bearing of points A and B from the point O.

i First measure the distances OA and OB.

OA is 5 cm and OB is 3.5 cm.

This means that OA is 50 km and OB is 35 km.

ii Draw a North line and place a protractor with its centre at O and 0° along the North line.

Measure clockwise to find the bearing.
A is on a bearing of 100°.
B is on a bearing of 240°.

Scale 1 cm : 10 km

O
A
B

N
O
A
B

Top Tip!

Always start at **North** as zero degrees and measure **clockwise**. Bearings under 100° should be written with a zero in front e.g. 090°.

Top Tip!

A full round protractor is better for measuring bearings.

Example: The point C is south of A and west of B.
Mark the position of C on the diagram.

A
B

Draw a line south from A and west from B.
Where the lines intersect is the position of C.

A
C
B

level 6

Back bearings

- If the bearing of the point B from A is 045°, then the bearing from B to A, called the **back bearing**, is 225°. (See the red line on diagram.)
- You can see this quite easily by looking at the angles opposite each other on a full round protractor.
  For example, 300° is opposite 120°.
  (See the green line on diagram.)
  Can you see the connection?

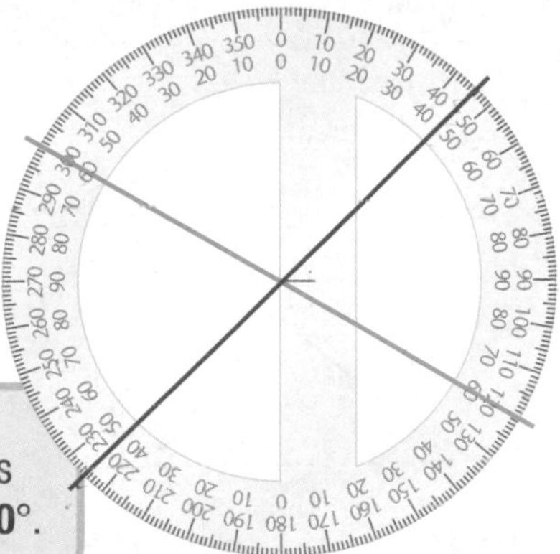

Top Tip!

The connection between a bearing and its back bearing is that the **difference** is **180°**.

Sample mental test question

Look at the diagram.
Estimate the angle marked.

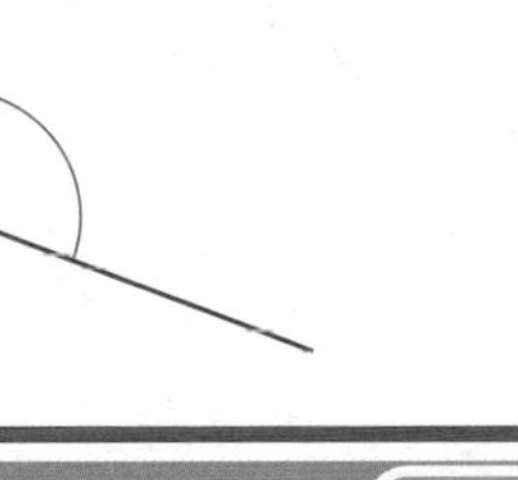

The angle is about 110°, so any answer from 100° to 120° would be an acceptable estimate.

level 6

Sample worked test question

The diagram shows two towns Alton and Beeton.

a Use the scale to find the actual distance of A from B.

b Find the bearing of A from B.

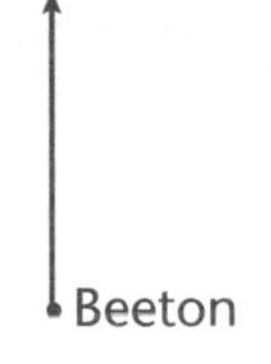

Scale: 1 cm represents 10 km

Answers

a *The measured distance from Alton to Beeton is 5 cm so the actual distance is 50 km.*

b *Draw a North line at B and measure clockwise to A. The bearing is 050°.*

Did You Know?

The magnetic North Pole and the geographic North Pole are not the same place. They are about 3° apart if you are in Britain.

Spot Check

1 What bearing is opposite to
a east **b** south **c** west **d** 130° **e** 200°?

SHAPE, SPACE AND MEASURES

Angle facts

level 5

Polygons

- You need to know some angle facts about **polygons**.

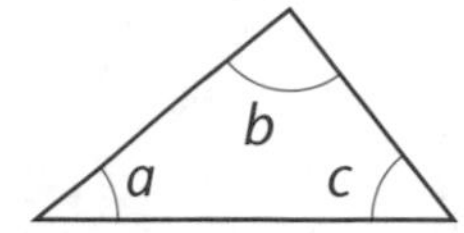

The **three interior** angles in a **triangle** add up to **180°**.

$a + b + c =$ **180°**

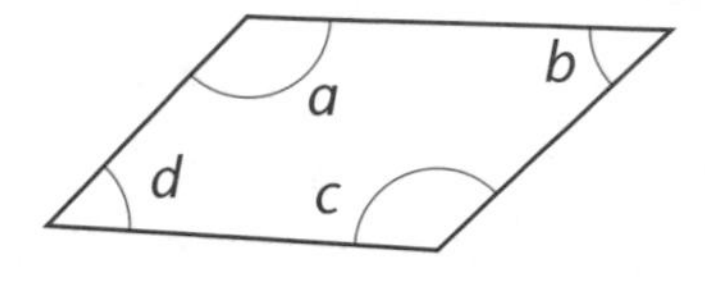

The **four interior** angles in a **quadrilateral** add up to **360°**.

$a + b + c + d =$ **360°**

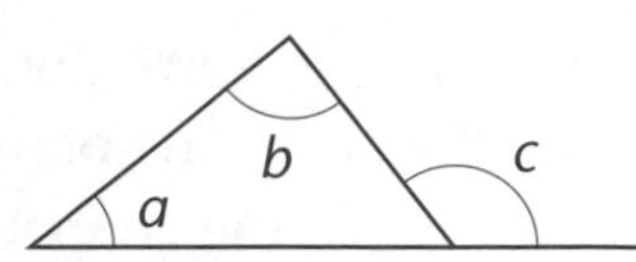

The **exterior** angle of a triangle equals the **sum** of the **two opposite interior** angles.

$a + b = c$

level 6

Special triangles

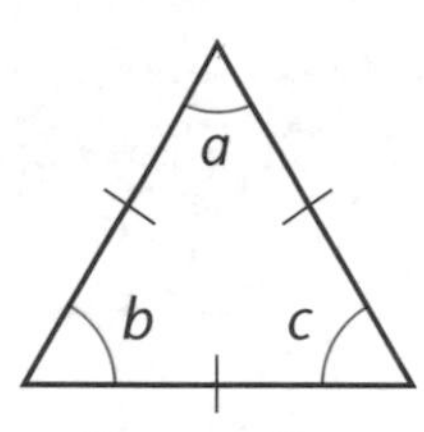

Equilateral triangle

$a = b = c =$ **60°**

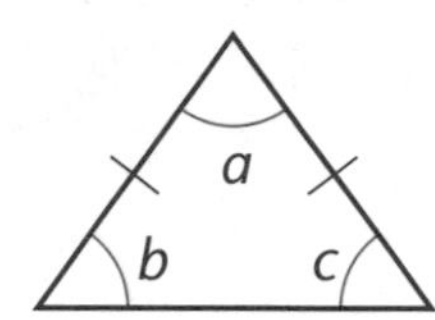

Isosceles triangle

$b = c$

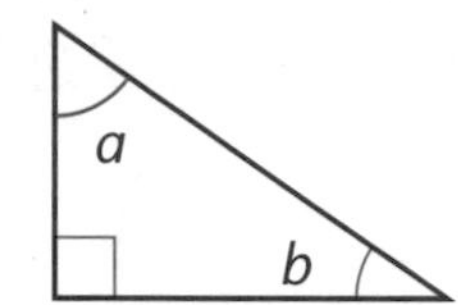

Right-angled triangle

$a + b =$ **90°**

Top Tip!

The small dashes marked on lines show that those lines are of the same length

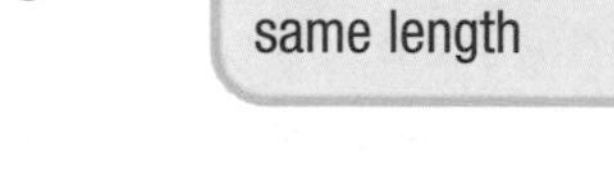

Example: Find the angles marked with letters in these shapes.

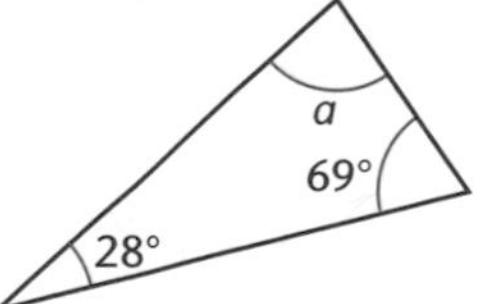

$a + 28 + 69 = 180$

$a + 97 = 180$

$a = 180 - 97$

$a = 83°$

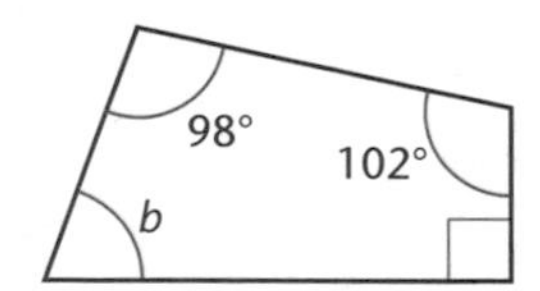

$b + 98 + 102 + 90 = 360$

$b + 290 = 360$

$b = 360 - 290$

$b = 70°$

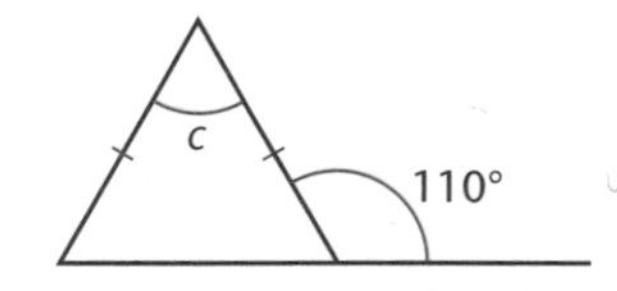

angle on base $= 180 - 110$

$= 70$

$c + 70 + 70 = 180$

$c = 40°$

1 What is angle x in this diagram?

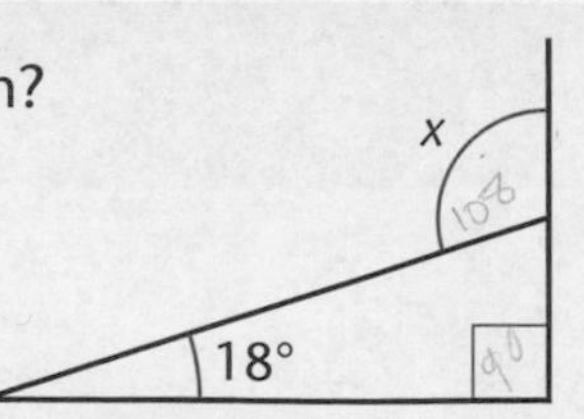

level 6

Interior and exterior angles

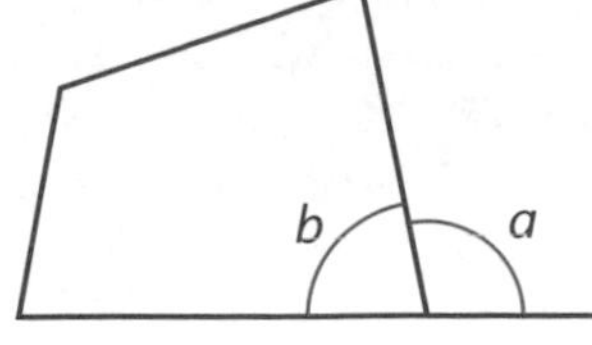

- When a side of a polygon is extended, the **angle formed** by the **extended side** and the **polygon** is called the **exterior angle**.
- In the diagram, a is the **exterior angle** and b is the **interior angle**.

Example: Find the angles in the following diagrams.

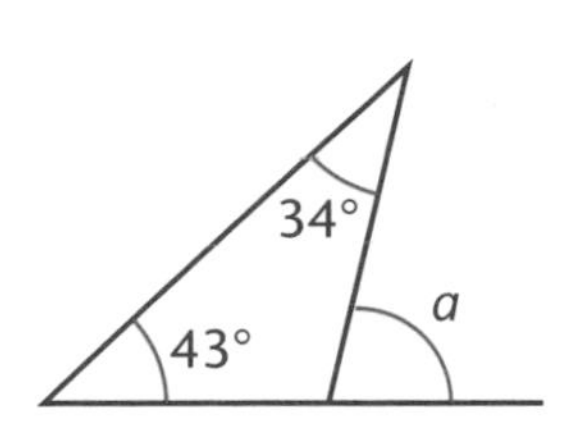

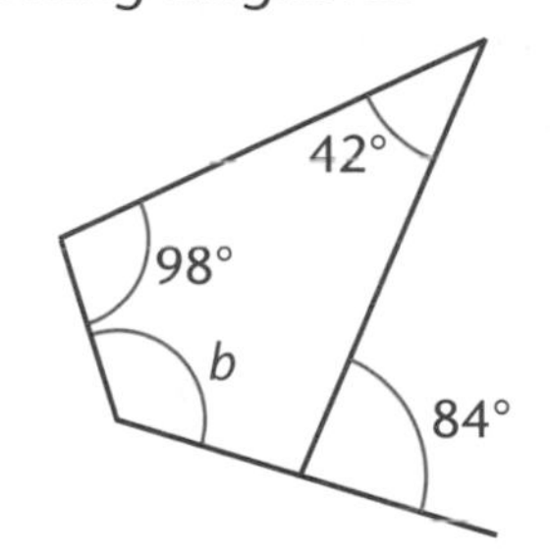

Top Tip!

In a triangle, any exterior angle is equal to the sum of the two opposite interior angles.

a The missing angle in the triangle is 180 – (34 + 43) = 180 – 77 = 103°.
The angle a on the straight line is 180 – 103 = 77°.

b The missing angle on the straight line is 180 – 84 = 96°.
Angle b is 360 – (98 + 42 + 96) = 360 – 236 = 124°.

Sample mental test question

Two angles in a triangle are 45° and 85°. How many degrees is the other angle?

Work out 45 + 85 and then take the answer away from 180.

45 + 85 = 130

180 – 130 = 50

The angle is 50°.

level 6

Sample worked test question

This shape has been made from two congruent isosceles triangles.
What is the size of angle a?

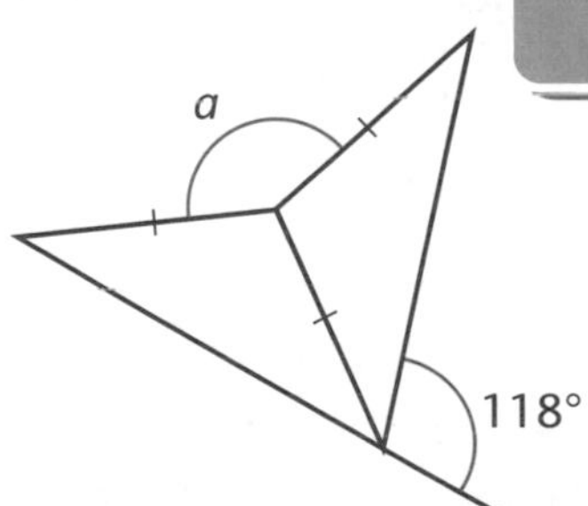

Answer
The angle on the same straight line as 118° is 180 – 118 = 62°.
This is double the angle in each isosceles triangle so these angles are 62 ÷ 2 = 31°.
This means the larger angle in the isosceles triangle is 180 – (2 x 31) = 118°
So, a = 360 – (2 x 118) = 360 – 236 = 124°.

Did You Know?

When mud dries in the sun, the cracks form curves that intersect at right angles.

SHAPE, SPACE AND MEASURES

Angles in parallel lines and polygons

level 6

Angles in intersecting lines and parallel lines

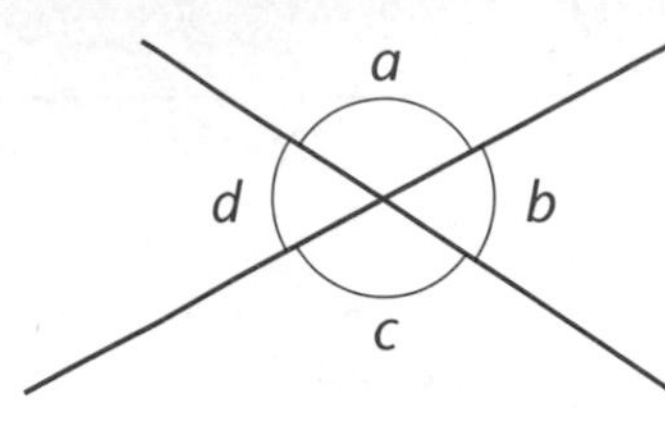

In these intersecting lines, vertically opposite angles are equal.

$a = c$ and $b = d$

- Parallel lines never meet. They provide some special types of angles.

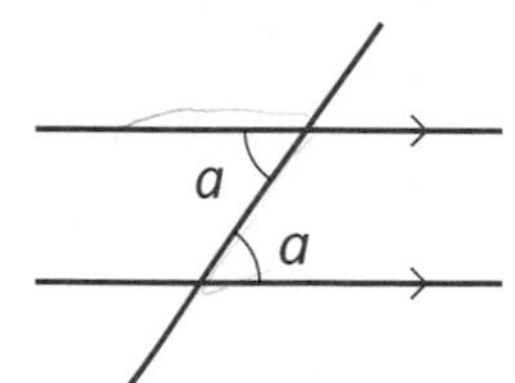

Alternate angles are **equal**.

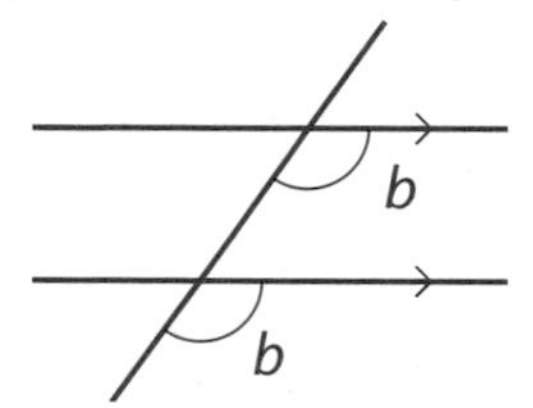

Corresponding angles are **equal**.

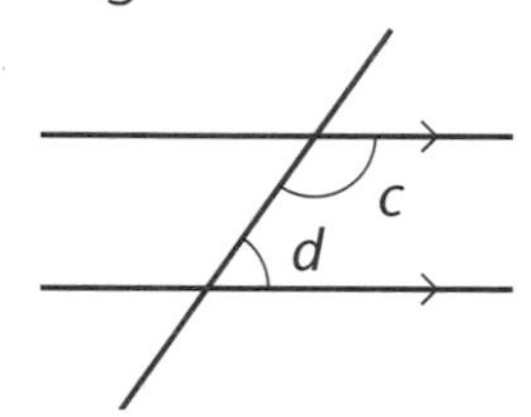

Interior angles add up to **180°**.

$c + d =$ **180°**

Example: Find the angles marked by letters.

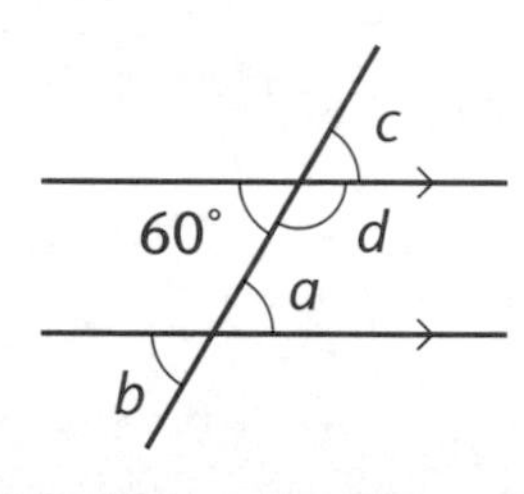

a = 60° (alternate angle)

b = 60° (vertically opposite angle to a)

c = 60° (corresponding angle to a)

d = 120° (interior angle to a)

level 5

Polygons

- Polygons are two-dimensional (**2-D**) shapes with **straight** sides.

Name of polygon	Number of sides	Sum of interior angles
Triangle	3	180°
Quadrilateral	4	360°
Pentagon	5	540°
Hexagon	6	720°
Heptagon	7	900°
Octagon	8	1080°

- A pentagon can be split into three triangles, so the sum of the five interior angles is 3 x 180° = 540°.

When solving angle problems, always give a reason for how you found each angle.
Always use the correct mathematical words:

- alternate angles
- corresponding angles
- interior angles.

level 6

Regular polygons

- **Regular polygons** are polygons with **all sides equal** and **all angles equal**.

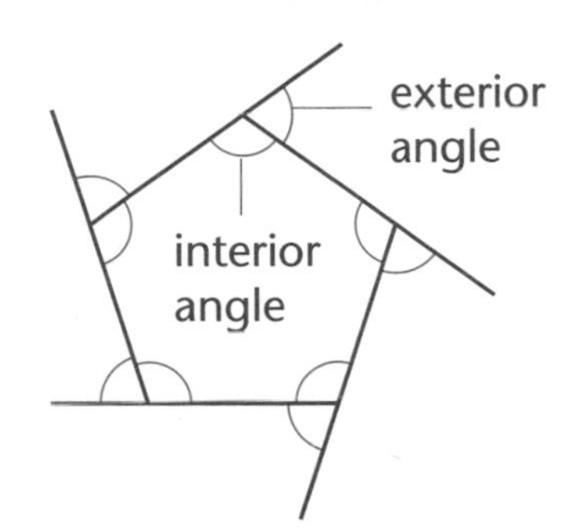

The regular pentagon has five equal interior angles and five equal exterior angles.

Sum of the five exterior angles = 360°

So each exterior angle = 72°

Interior angle + exterior angle = 180°

Each interior angle = 108°

level 6

Sample worked test question

ABCD is a rectangle.

Find the angles marked with a letter.

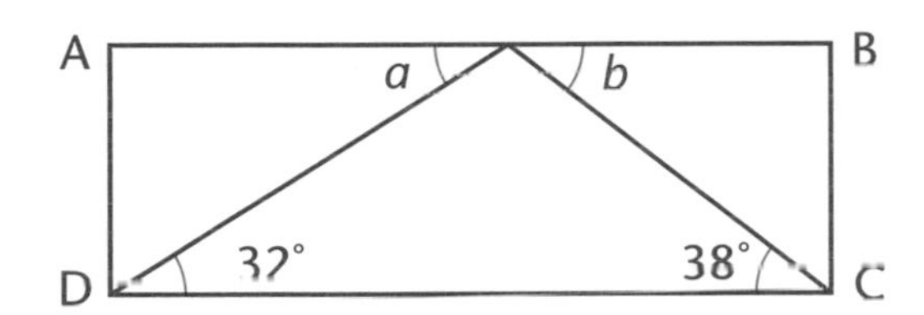

Answers

AB is parallel to CD, so a = 32° (alternate angle) and b = 38° (alternate angle).

Top Tip!

Alternate angles: look for a 'Z'.

Corresponding angles: look for an 'F'.

Interior angles: look for a 'C'.

Islamic art is based on polygons, which can be constructed with circles.

1 What are the missing angles in this regular hexagon?

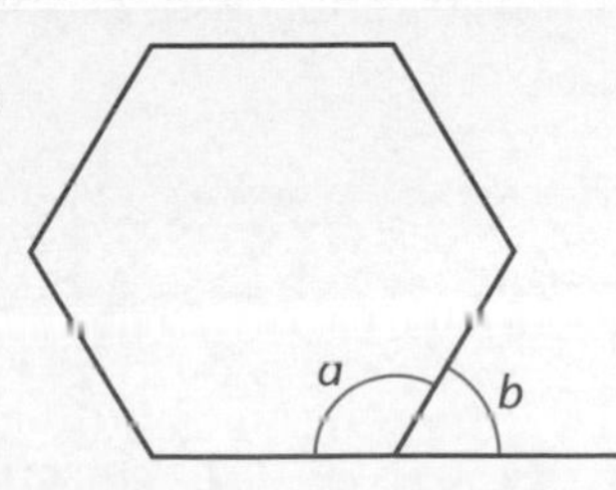

2 Find the missing angles in this diagram.

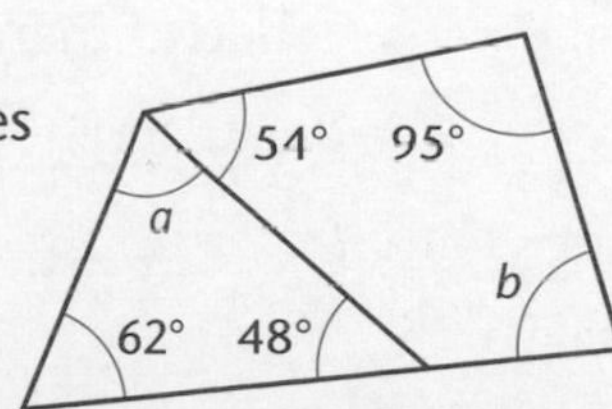

Reflections and rotations

level 6

Reflection

- A **reflection** is a transformation that **reflects** a shape in a **line of symmetry** (or a mirror line).

Example: Reflect these shapes in the mirror lines shown.

a

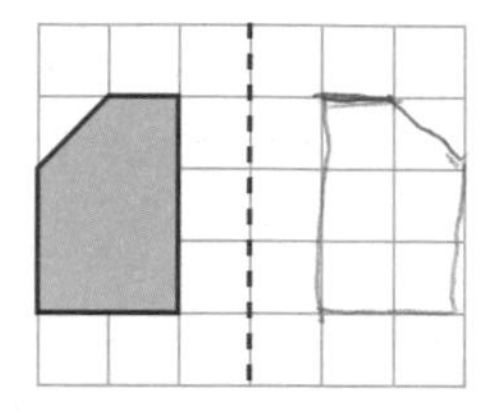

b

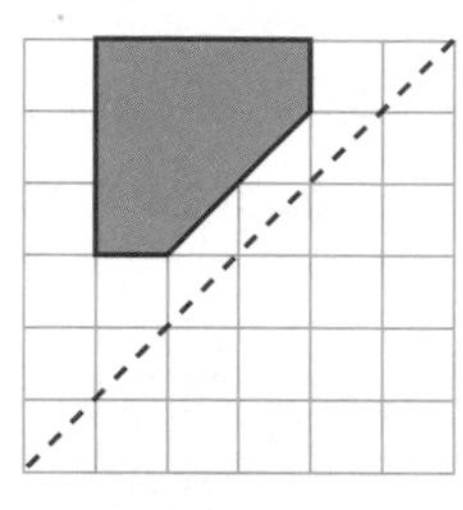

a

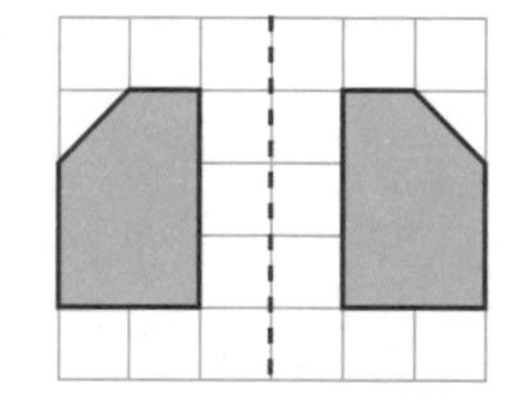

b

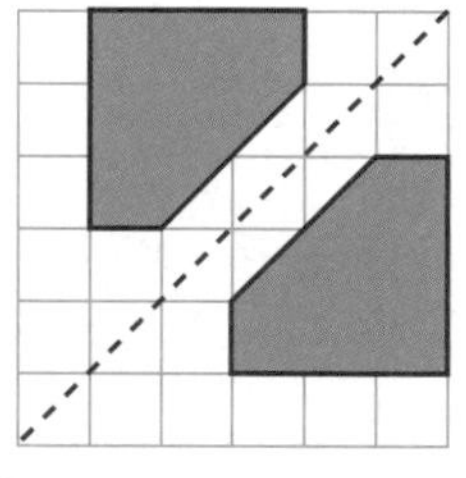

Example: Reflect the triangle shown in

a the line $y = -1$

b the line $y = -x$.

Top Tip!

If the line of symmetry is not vertical or horizontal, turn the page round so that it is.

First identify the mirror lines, then reflect the triangle.

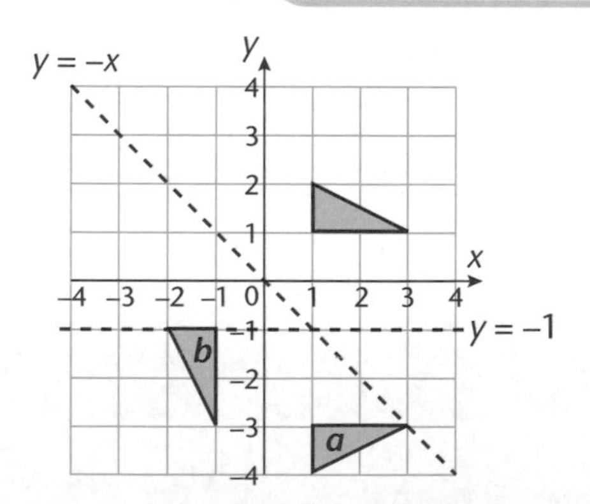

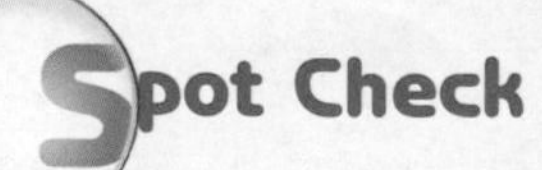

1 Describe the transformations that take the shaded triangle to

a triangle A **b** triangle B.

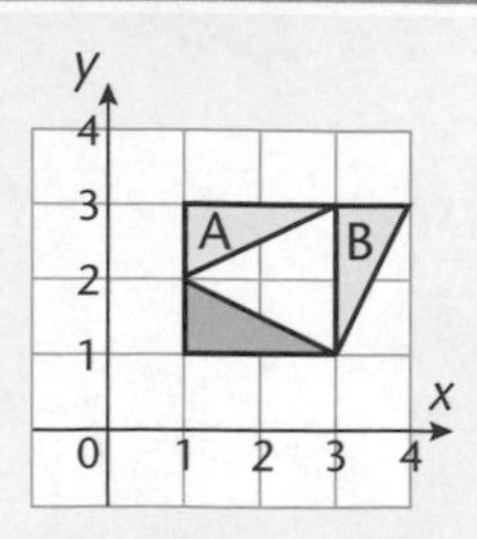

level 6

Rotation

- A **rotation** is a transformation that **rotates** a shape, in either a **clockwise** or **anticlockwise** direction, through a **given angle** about a **given centre**.

Example: Rotate these shapes by 90° clockwise about the centre O.

a

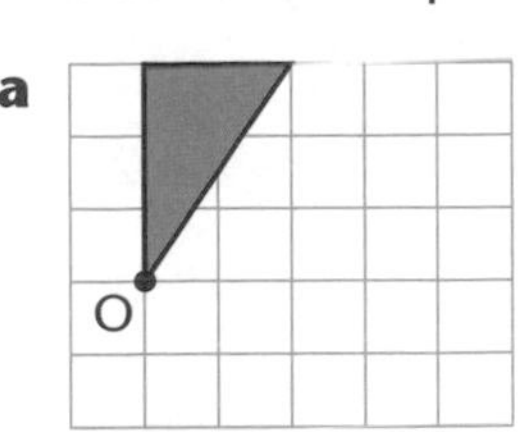

b

O

Top Tip!

Use tracing paper to copy the shape and a pencil point as the centre. Then rotate the shape using the tracing paper.

a

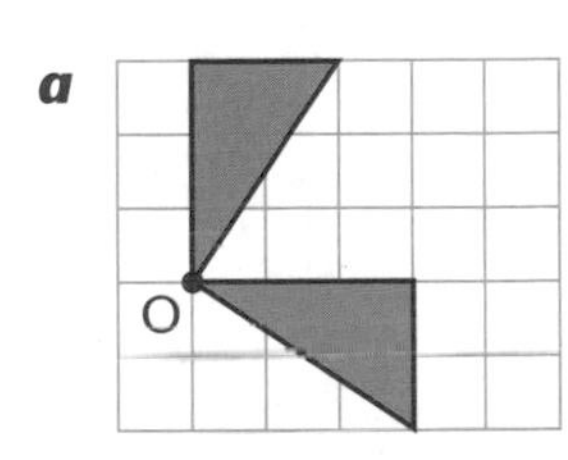

b

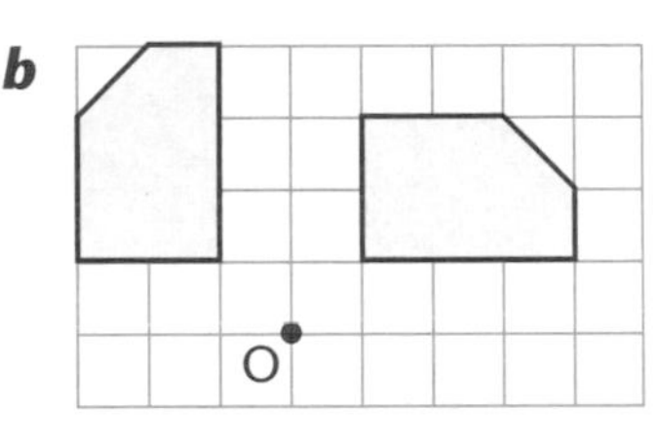

Example: Rotate the given triangle by

a 180° anticlockwise about (0, 2)

b 90° clockwise about (–1, 0).

First identify the centres, then rotate the triangle.

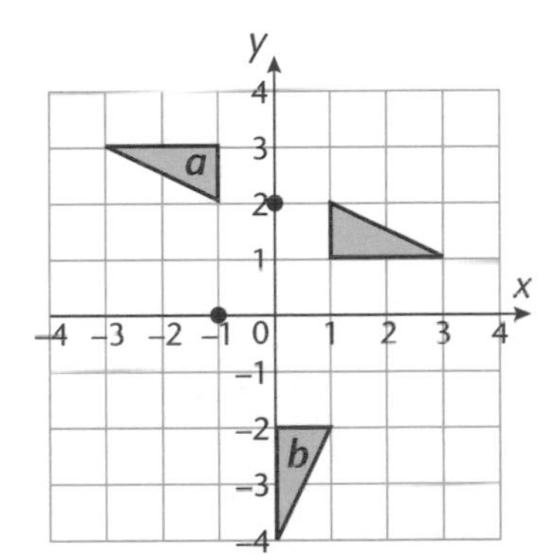

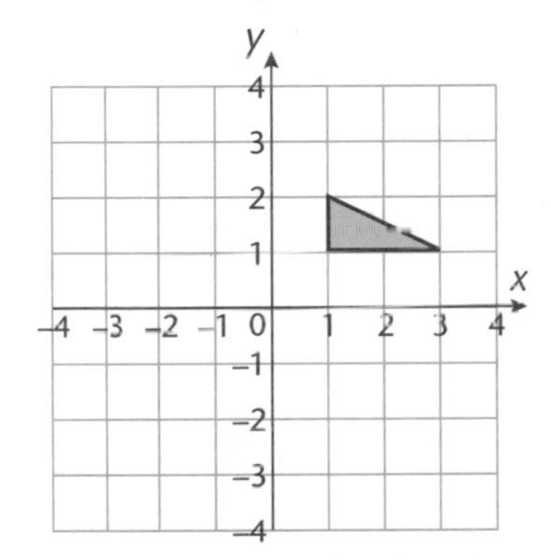

level 6

Sample worked test question

a The kite is transformed so that A is transformed to the point A′ and B is transformed to the point B′. Describe the transformation.

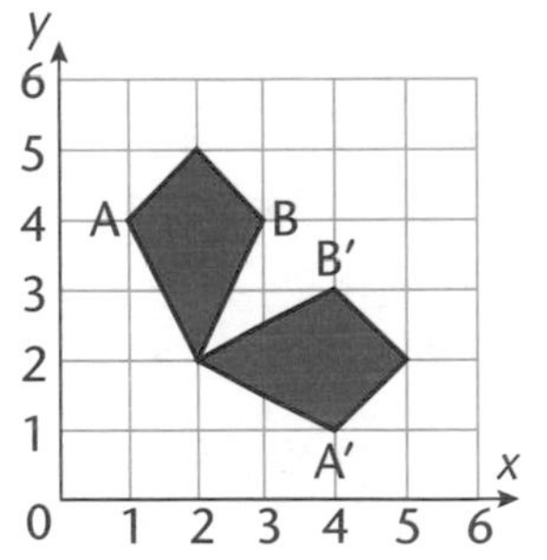

b The kite is transformed so that A is transformed to the point A′ and B is transformed to the point B′. Describe the transformation.

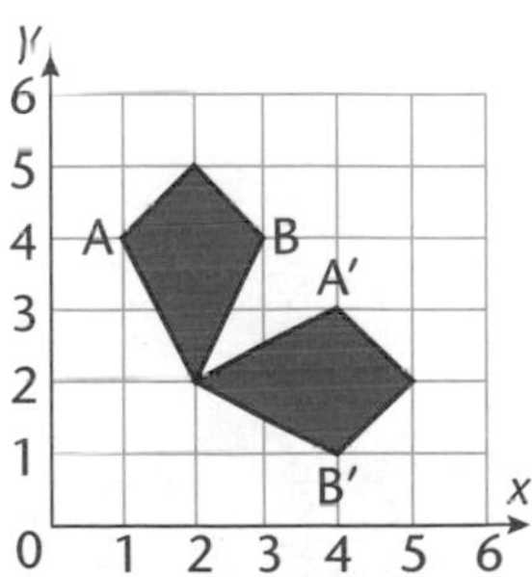

Answers

a *Reflection in the line $y = x$.*

b *90° clockwise rotation about the point (2, 2).*

Did You Know?

The Earth rotates once every day. Venus rotates once every 243 Earth days. Nights on Venus can be very long!

SHAPE, SPACE AND MEASURES

Enlargements

Scale factor

level 6

- An **enlargement** changes the **size** of a **shape**.
- The **scale factor** tells you **how many times bigger** the shape is to be enlarged.
- The shape will stay the same but the sides will all increase by the same factor.

Example: Shape A is enlarged to shapes B and C. What are the scale factors of the enlargements?

Compare the lengths of any two common sides.
Shape B has a scale factor of $1\frac{1}{2}$.
Shape C has a scale factor of 2.

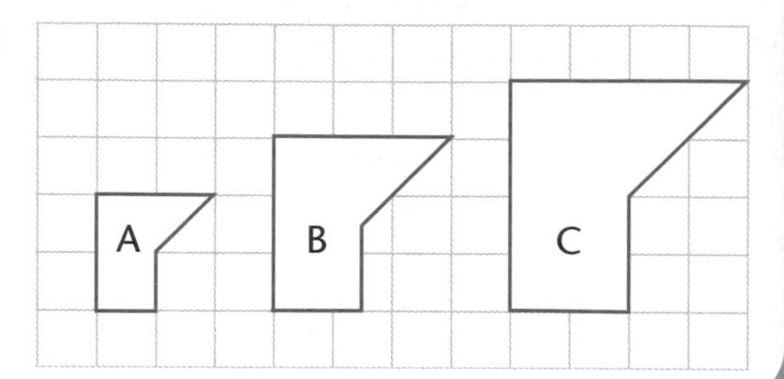

Centre of enlargement

level 6

- To enlarge a shape, you also need a **centre of enlargement**.

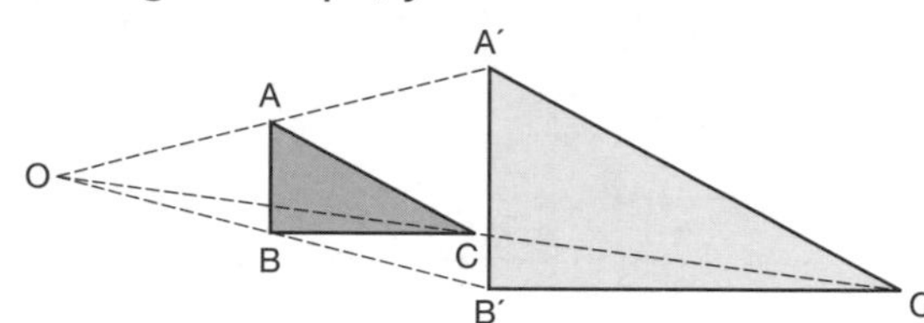

Triangle ABC is enlarged by a scale factor of 2 about the centre of enlargement O.
All the sides are doubled in length.

OA′ = 2 x OA

OB′ = 2 x OB

OC′ = 2 x OC

Notice that all the measurements are from O.

Example: Enlarge the triangle ABC about the origin O by a scale factor of 3.

The coordinates of the vertices of triangle ABC are: A(3, 2), B(3, 1) and C(1, 1).

The coordinates of the vertices of triangle A′B′C′ are: A′(9, 6), B′(9, 3) and C′(3, 3).

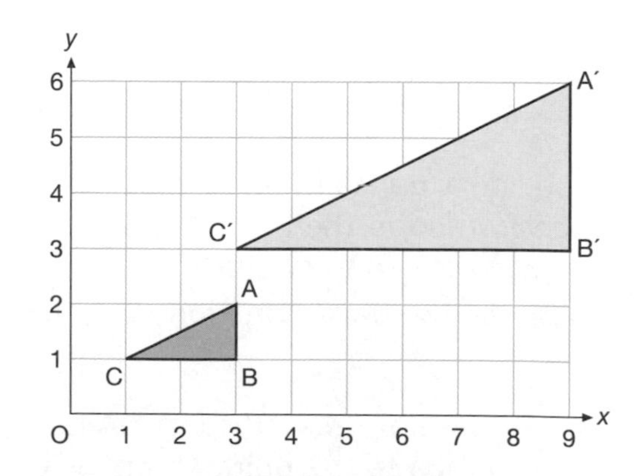

Top Tip!

If you draw lines through common vertices, they will meet at the centre of enlargement.

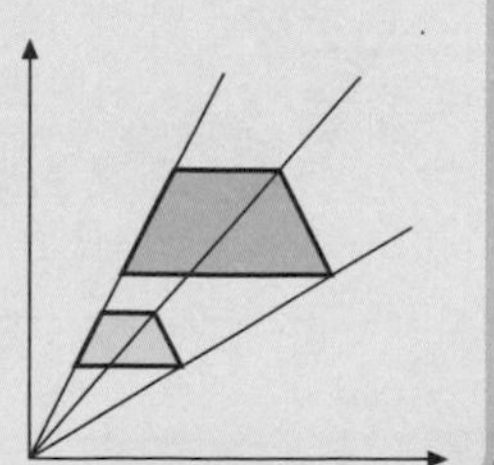

Top Tip!

The coordinates of the vertices of triangle ABC above are multiplied by the scale factor to give the vertices of triangle A′B′C′. This method works if the centre of enlargement is at the **origin**, but not if it is elsewhere.

level 6

Sample worked test question

Enlarge the trapezium ABCD by a scale factor of 2 about the origin O.

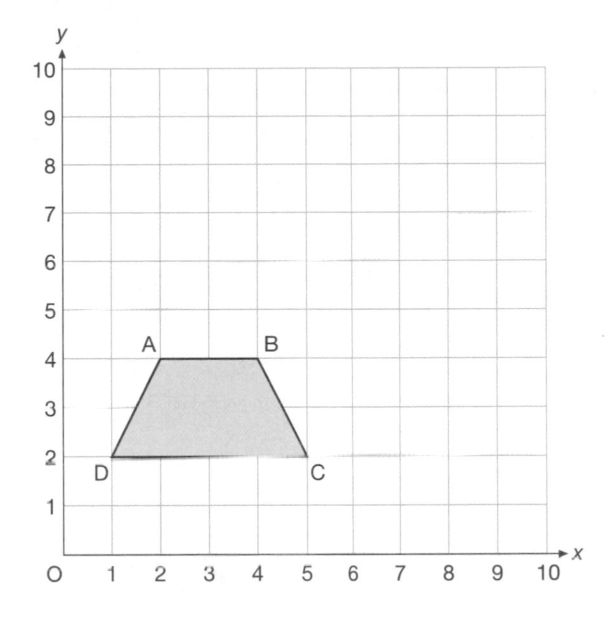

Answer

The coordinates of the enlarged trapezium A′B′C′D′ are:
A′(4, 8), B′(8, 8), C′(10, 4) and D′(2, 4).

Did You Know?

If you place a sheet of paper on the floor and keep doubling the size of the pile, i.e. 1 sheet becomes 2, 2 sheets becomes 4 etc., then after 50 doubles the pile will be 100 million kilometres high.

Spot Check

1 Enlarge this triangle by a scale factor of 3.

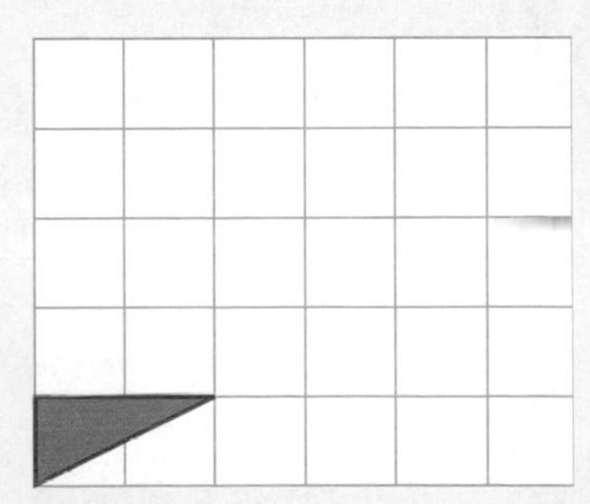

2 The sides of a rectangle are doubled in size. Explain why the area of the new rectangle is not twice as big as the area of the original.

3-D shapes

level 5

Polyhedra

- **2-D** shapes are called **polygons** and **3-D** shapes are called **polyhedra**.
- These are the names of the 3-D shapes you need to know.

Cube

Cuboid

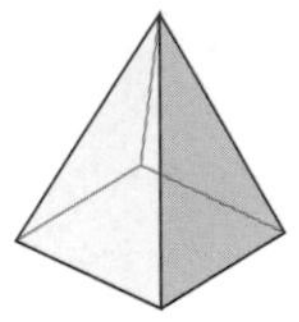
Square-based pyramid

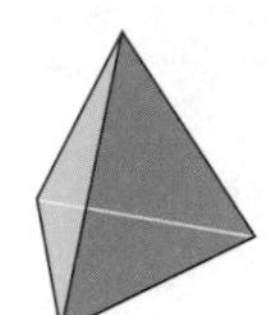
Tetrahedron

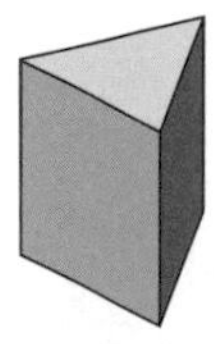
Triangular prism

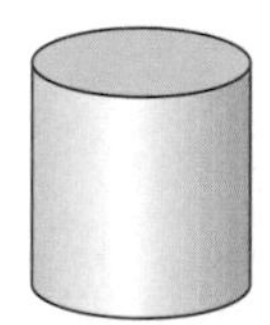
Cylinder

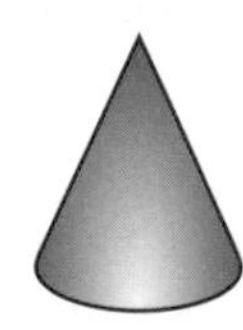
Cone

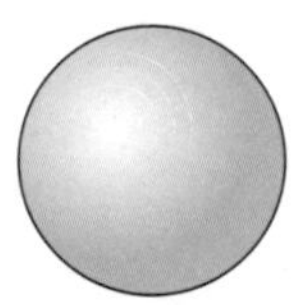
Sphere

- A **cuboid** has 12 **edges**, 8 **vertices** and 6 **faces**.

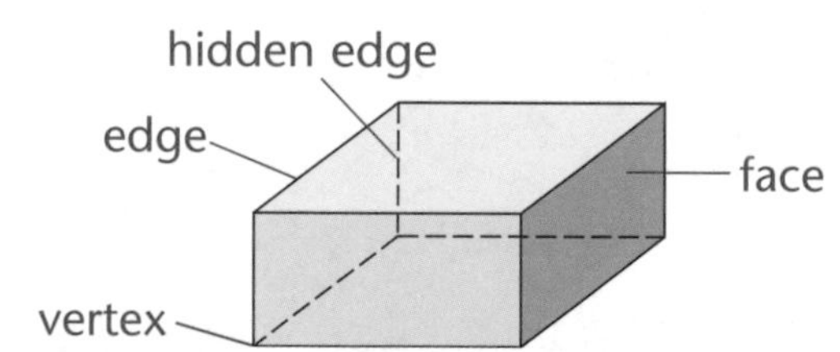

level 5

Nets

- A **net** is a 2-D shape that can be folded to make a 3-D shape.

Example: This is the net of a cylinder.

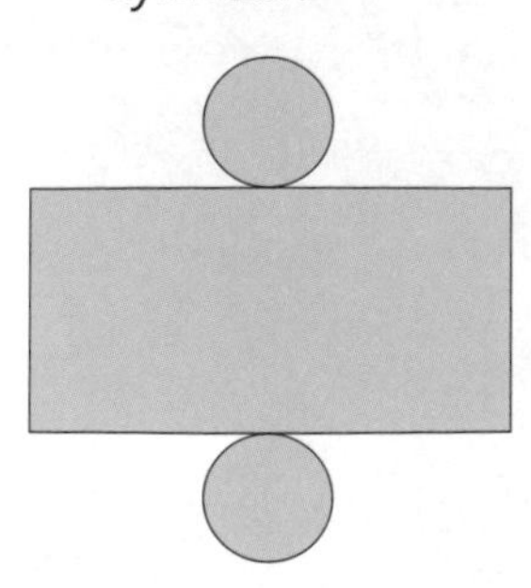

The dimensions of the rectangle are $2\pi r$ by h, and each circle has a radius of r.

level 5

Plans and elevations

- A **plan** is a view of a 3-D shape **from above**.
- An **elevation** is a view of a 3-D shape from **one side**.

Example: These are the views for a triangular prism.

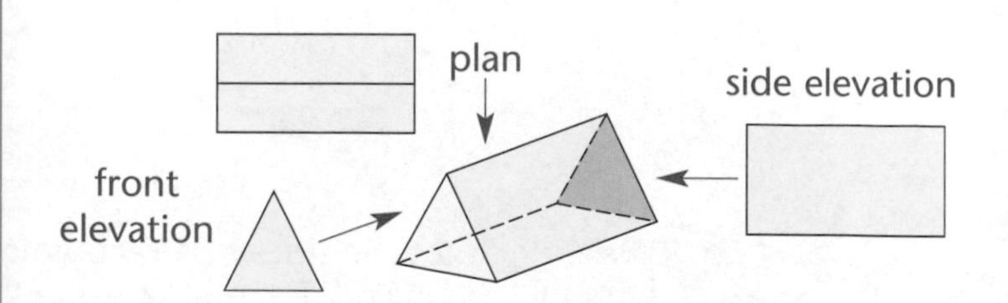

Isometric drawings

level 5

- 3-D shapes drawn on isometric paper are more accurate and measurements can be taken from the diagram.

When using isometric paper, the dots should form columns:

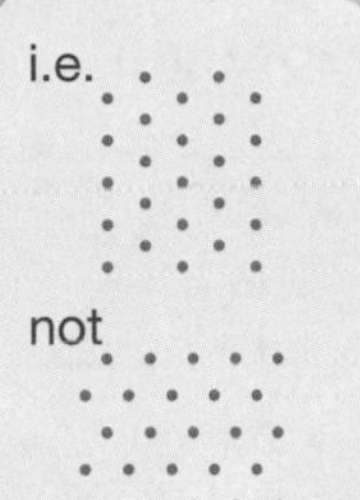

Example: This is the isometric drawing for a cuboid.

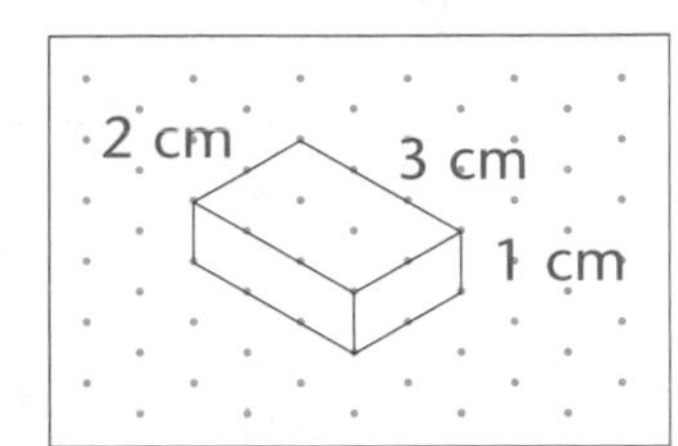

Planes of symmetry

level 5

- A 3-D shape has **plane symmetry** if it can be **cut in half** so that one half is a **mirror image** of the other half.

Example: A cuboid has three planes of symmetry.

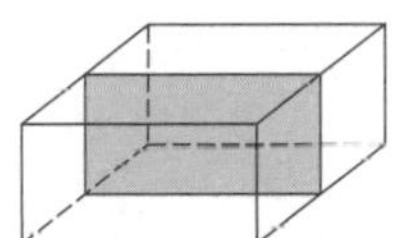

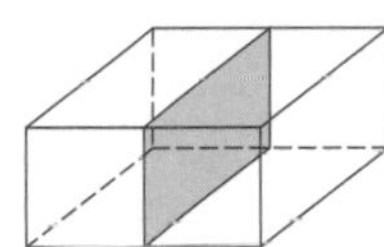

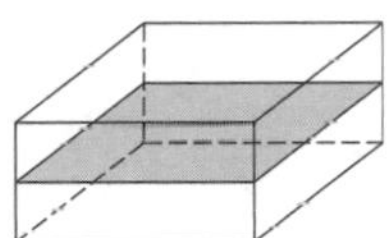

Sample worked test question

level 5

The diagram shows a shape made from 5 one centimetre cubes.
On the grid below draw **a** the plan **b** the elevation from X.

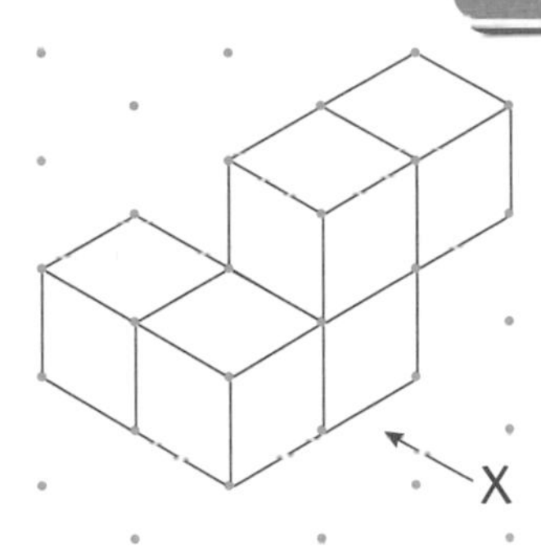

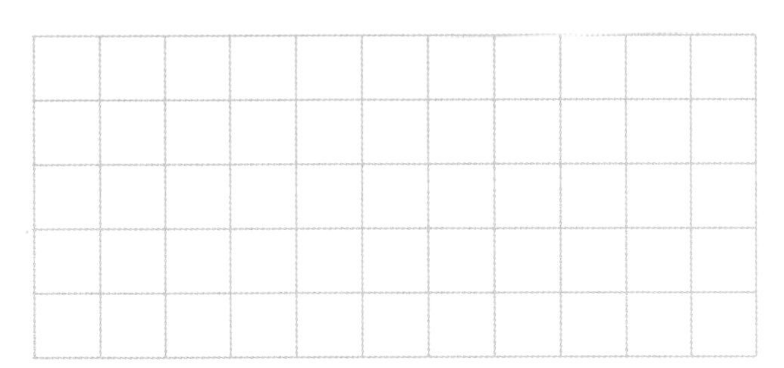

Answers

a ***b***

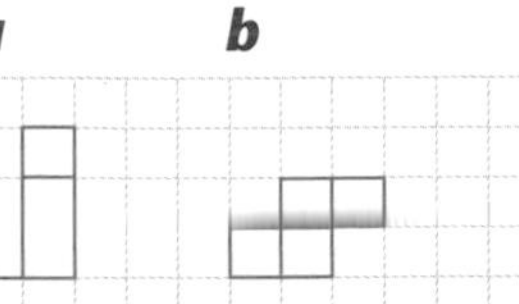

Did You Know?

Elevation is another term for height above sea level. Lake Titicaca is the highest lake in the world with an elevation of 3810 m. The Dead Sea is the lowest with an elevation of –411 m.

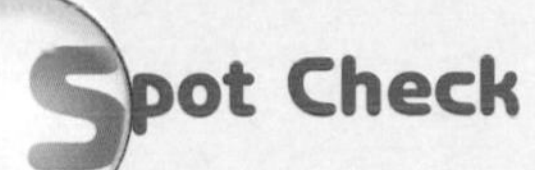

1 Which of the following are nets for a cube?

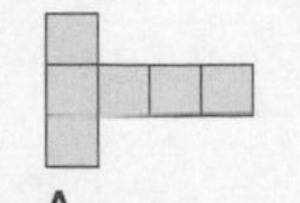

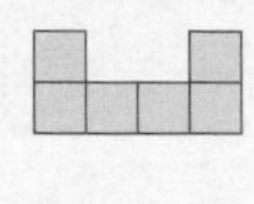

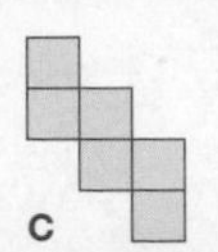

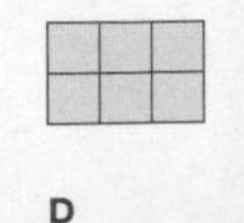

Perimeter and area

Area formulae

levels 5-6

- In these formulae, b stands for **base** and h stands for **height** although the correct term is **perpendicular height**.

Parallelogram

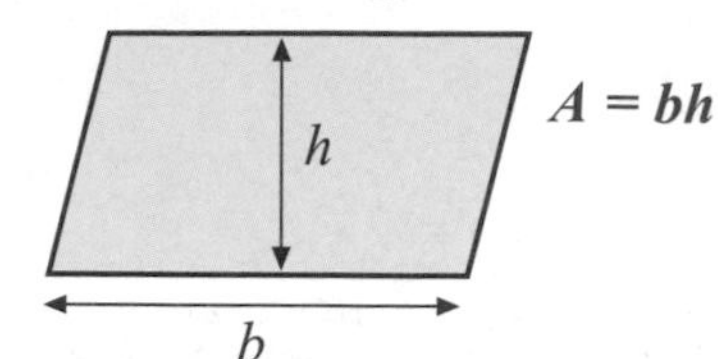

Triangle

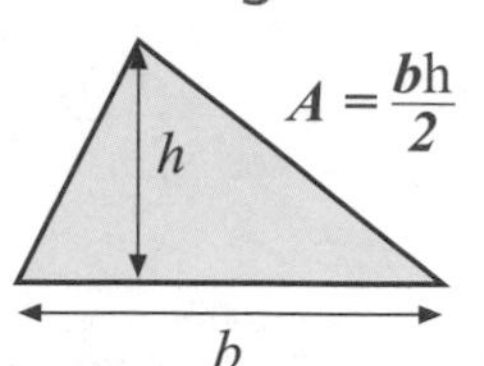

Trapezium

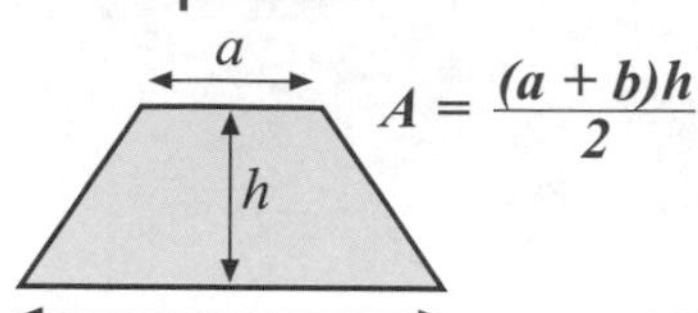

$A = bh$ $\quad A = \frac{bh}{2}$ $\quad A = \frac{(a+b)h}{2}$

Example: Find the area of these shapes.

a

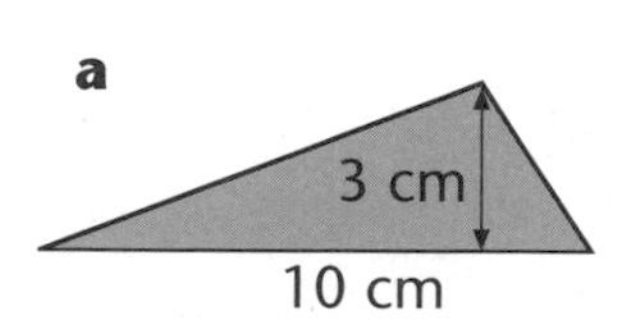

b

5 cm
4 cm
9 cm

a $A = \frac{bh}{2} = \frac{10 \times 3}{2} = 15 \text{ cm}^2$ ***b*** $A = \frac{(a+b)h}{2} = \frac{(9+5) \times 4}{2} = 28 \text{ cm}^2$

Top Tip!

Do not try to work these out in your head. Always substitute numbers into the formula. You will have more chance of reaching the correct answer.

Area of combined shapes

levels 5-6

- Shapes can be made up of **combinations** of rectangles, triangles etc.
- To work out the area of these shapes, break them down into shapes that you know the formulae for like triangles, rectangles, parallelograms and trapezia.

Example: Find **a** the perimeter and **b** the area of this shape.

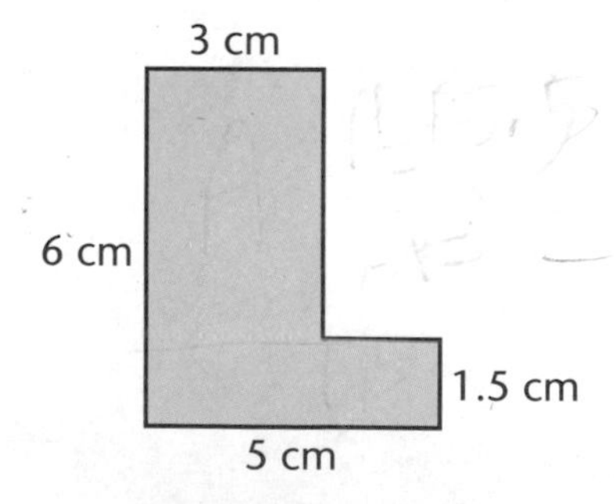

Split the shape into two rectangles and label them A and B.

Work out the sides of each of the rectangles.

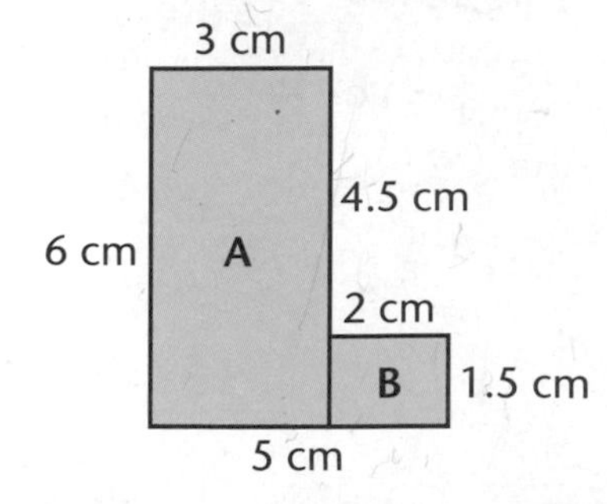

a Perimeter = 6 + 3 + 4.5 + 2 + 1.5 + 5 = 22 cm

b Area A $= 3 \times 6 = 18 \text{ cm}^2$
Area B $= 2 \times 1.5 = 3 \text{ cm}^2$ +
Total $= 21 \text{ cm}^2$

Area of combined shapes continued

Example: Find the area of this shape.

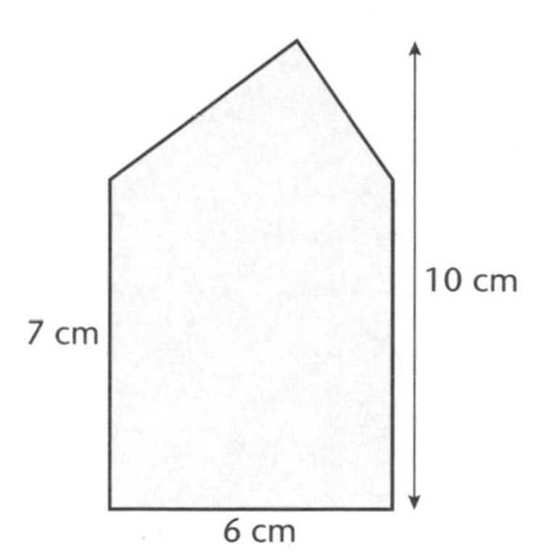

Split the shape into a rectangle and a triangle.

Work out the height of the triangle.

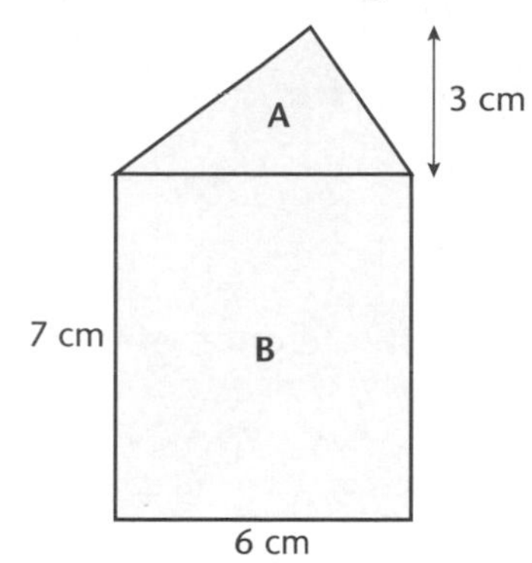

Then work out the area of each shape.

Area A	$= \frac{1}{2} \times 6 \times 3$	$= 9\ cm^2$
Area B	$= 7 \times 6$	$= 42\ cm^2$ +
Total		$= 51\ cm^2$

Sample mental test question

A square has an area of 36 cm^2. What is the perimeter of the square?

The length of a side of the square is 6 cm, since 6 x 6 = 36.
So, the perimeter of the square is 4 x 6 = 24 cm.

Did You Know?

On average there are about 4000 square metres of land for every person in Britain. That's less than a soccer pitch.

Sample worked test question

level 6

A shape is made from two triangles A and B and a rectangle C.
The triangle A has an area of 12 cm^2. Work out the area of the shape.

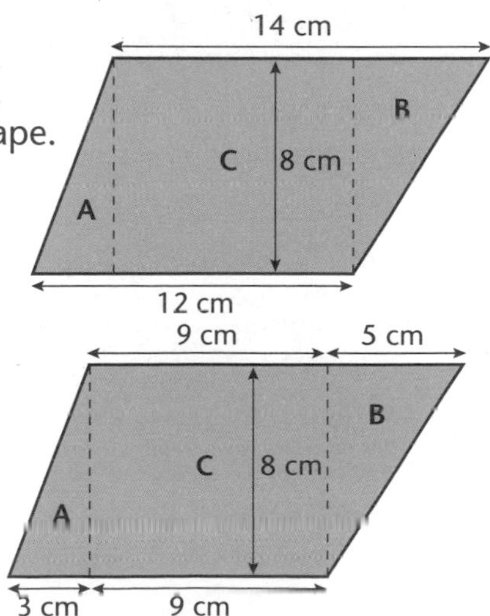

Answer

The area of triangle A is given by the formula $A = \frac{bh}{2}$

If the height is 8 cm and the area is 12 cm^2,

$12 = \frac{1}{2} \times b \times 8$ *so the base of A must be 3 cm.*

This means the shape can be split up as

Area A =	=	*12* cm^2
Area B = $\frac{1}{2} \times 5 \times 8$	=	*20* cm^2
Area C = 8 x 9	=	*72* cm^2 +
Total		*104* cm^2

Spot Check

1 What is the perimeter and area of this shape?

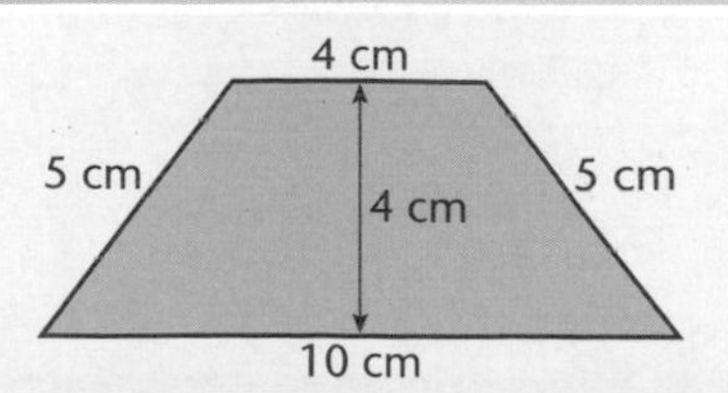

Circumference and area of a circle

level 6

Circumference of a circle

- The **circumference** is the **perimeter** of a circle.
- You need to know that $d = 2 \times r$.

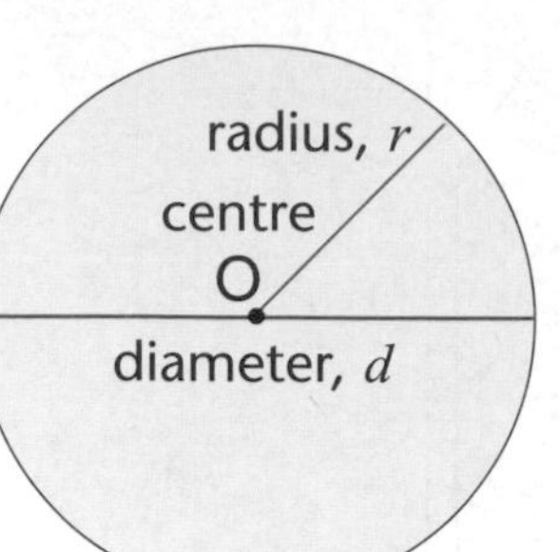

There are two formulae for the circumference of a circle:

- Circumference = 2 x π x radius

  $C = 2\pi r$
- Circumference = π x diameter

  $C = \pi d$
- π is a decimal that goes on forever. Take π as 3.14 to 2 decimal places or use the π key on your calculator.

Top Tip! Always give your answer to 1 decimal place (1 d.p.) unless the question says otherwise.

Example: Calculate the circumference of this circle.

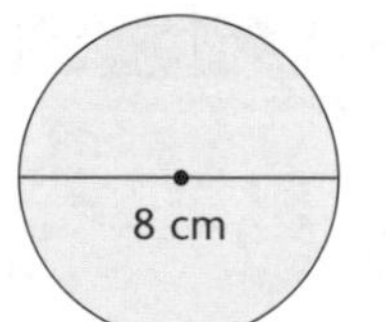

$C = \pi d = \pi \times 8 = 25.1$ cm (1 d.p.)

Key sequence on your calculator: π

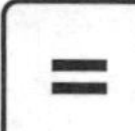

level 6

Area of a circle

- Area = π x radius²

  $A = \pi \times r \times r$

  $A = \pi r^2$

Top Tip! Questions will give either the radius or diameter. Make sure you use the correct value in the appropriate formula.

Example: Calculate the area of this circle.

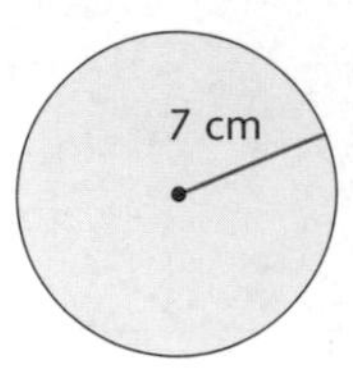

Top Tip! Write down the formula first and always show your working.

Always square the radius before you multiply by π.

$A = \pi r^2 = \pi \times 7^2 = 153.9$ cm² (1 d.p.)

Key sequence on your calculator: π

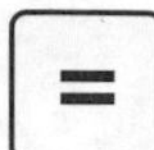

level 6

Sample worked test questions

a James' bike wheel has a radius of 30 cm. Calculate its circumference, giving your answer to the nearest centimetre.

b The circle and the square have the same area.

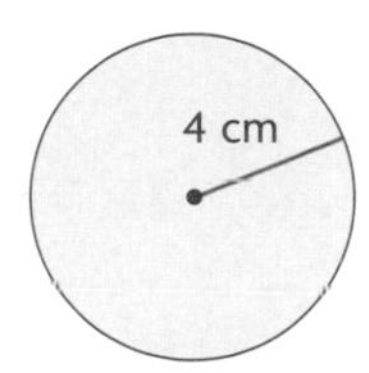

Calculate x, the length of the side of the square.

Answers

a $r = 30$ *cm , so* $d = 60$ *cm.*

$C = \pi d = \pi \times 60 = 188$ *cm (nearest cm)*

b *The area of the circle is* $A = \pi r^2 = \pi \times 4^2 = 50.26...$

So $x = \sqrt{50.26...} = 7.1$ *cm (1 d.p.)*

Did You Know?

In early 2006 Chris Lyons from Australia recited the first 4400 digits of pi from memory.

Spot Check

1 What is the circumference and area of this circle?

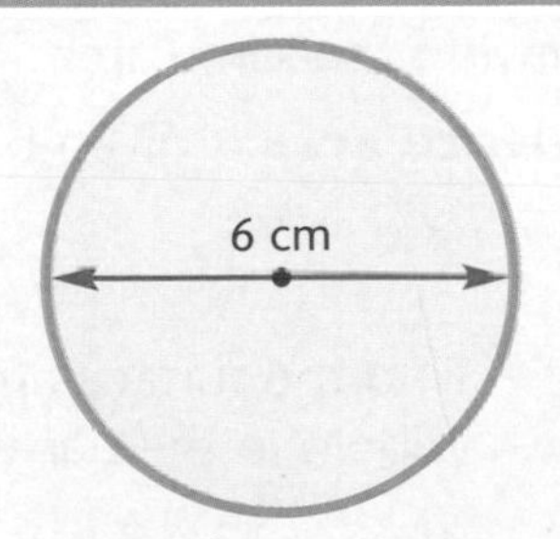

2 A circular pond has a diameter of 10 metres.
A one metre path is built around the outside of the pond.
What is the area of the path?

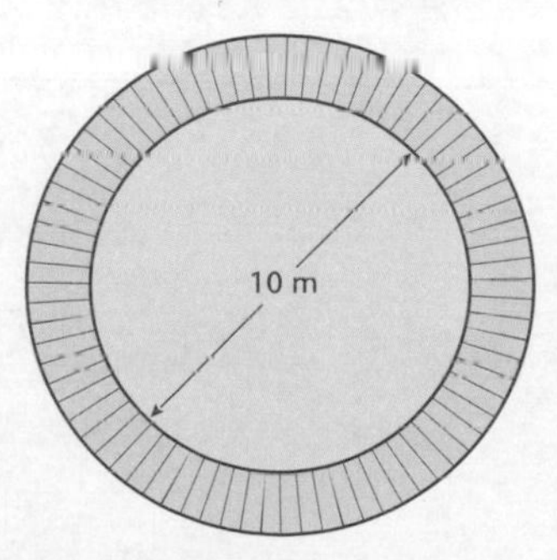

Volume

level 5

Volume of a cuboid

- **Volume** is the **amount of space** inside a **3-D** shape.
- The common units for volume are: mm^3, cm^3 or m^3.

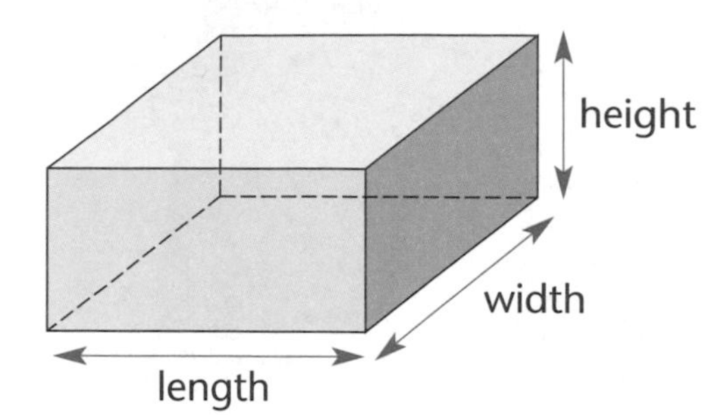

- Volume = length x width x height

  $V = l \times w \times h$

  $V = lwh$

Example: Find the volume of this cuboid.

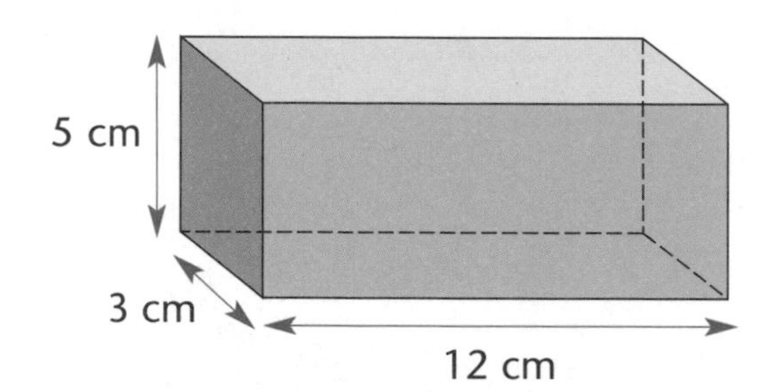

$V = lwh$

$= 12 \times 3 \times 5 = 180\ cm^3$.

Top Tip!

Substitute numbers into a formula before trying to work anything out.

level 6

Surface area of a cuboid

- There are 6 **faces** on a cuboid, with opposite faces having the same area.
- The **surface area** is given by

  $A = 2lw + 2lh + 2wh$

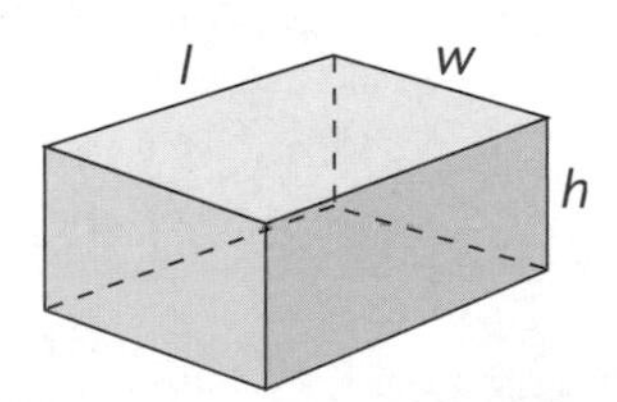

Example: Find the surface area of the purple cuboid in the panel above.

$A = 2 \times 12 \times 3 + 2 \times 12 \times 5 + 2 \times 3 \times 5 = 72 + 120 + 30 = 222\ cm^2$.

1 What is the volume and surface area of this cuboid?

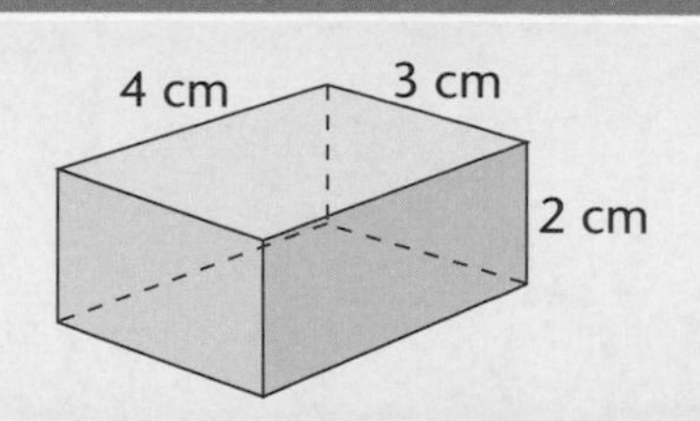

level 6

Capacity

- **Capacity** is the **amount of space** inside a **hollow 3-D** shape.
- Capacity usually refers to the volume of a gas or liquid. You need to know $1000 \text{ cm}^3 = 1$ litre.

Example: Find the volume of this fish tank, giving your answer in litres.

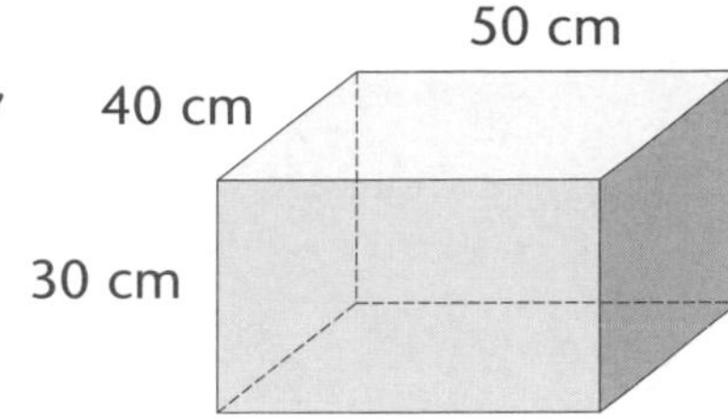

$V = 50 \times 40 \times 30$
$= 60\,000 \text{ cm}^3$
$V = 60$ litres

Example: This is a net of a cuboid. If one square has an area of 1 cm^2, what is the volume of the cuboid?

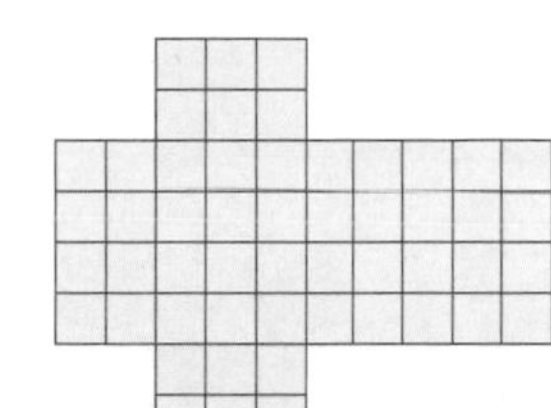

$V = 4 \times 3 \times 2 = 24 \text{ cm}^3$

Sample mental test question

The volume of a cube is 27 cm^3.
What is the length of an edge of the cube?

Since 27 = 3 x 3 x 3, the length of an edge = 3 cm.

Top Tip!

You should know these cube roots:

$\sqrt[3]{1} = 1$
$\sqrt[3]{8} = 2$
$\sqrt[3]{27} = 3$
$\sqrt[3]{64} = 4$
$\sqrt[3]{125} = 5$
$\sqrt[3]{1000} = 10$

level 6

Sample worked test question

These two cuboids have the same volume.
Find the value of x.

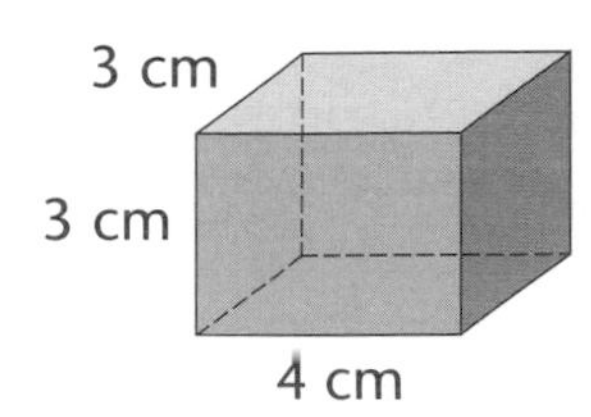

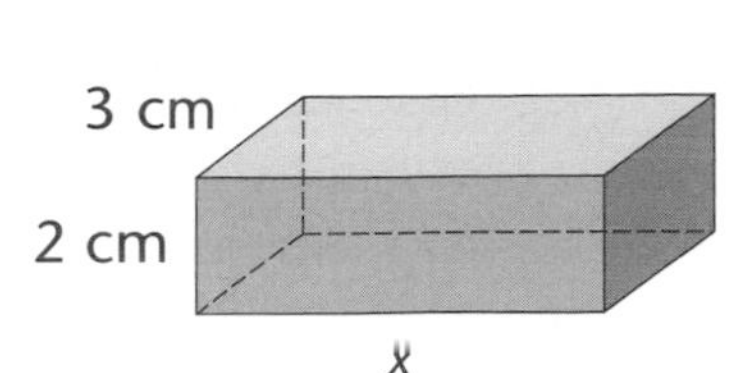

Answer
Volume of first cuboid = 36 cm^3.
So volume of second cuboid = $6x = 36 \text{ cm}^3$.
So $x = 6$ cm.

Did You Know?

The volume of the Sun can hold over a million Earths.

Constructions

Constructing triangles

level 6

- When constructing triangles it is very important that you **measure** lines and angles **accurately**.

Example: Construct this triangle accurately.

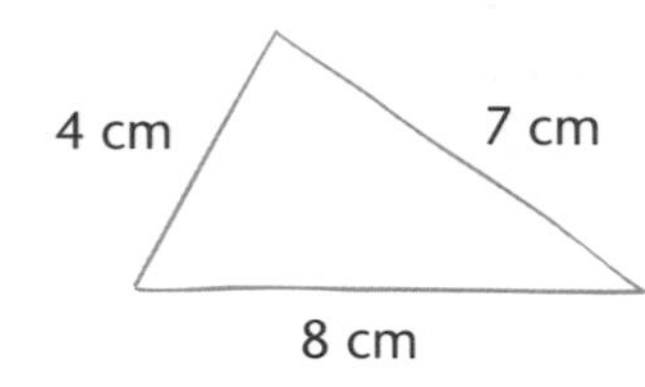

Part of the diagram, such as the base line, is often drawn for you. Always use a compass to mark out the distances rather than a ruler.

First draw a line 8 cm long.

Then use a compass to measure 4 cm and draw an arc from the left-hand end of the 8 cm line.

Then use a compass to measure 7 cm and draw an arc from the right-hand end of the 8 cm line.

Then join the ends of the 8 cm line to the point where the arcs cross.

Perpendicular bisector

level 7

- The **perpendicular bisector** is the **line** that passes through the **midpoint** of two other points and is **perpendicular** (at right angles) to the line that joins them.

Example: Construct the perpendicular bisector of AB.

A•   •B

First set the compass to about two-thirds of the distance from A to B.

Draw arcs from A on both sides of the line.

Without changing the size of the compass, do the same from B.

Join the points where the arcs cross.

This line is the perpendicular bisector of AB.

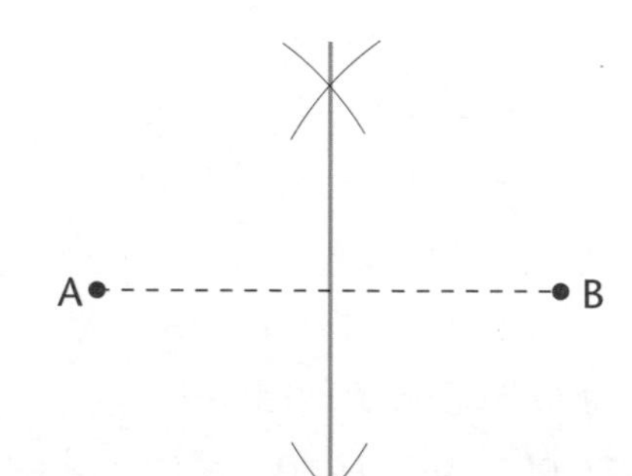

Top Tip!

Make sure your arcs are shown.

Angle bisector

level 7

- The **angle bisector** is the **line** that passes at the **same distance** from **two intersecting lines**.

Example: Construct the angle bisector of ABC.

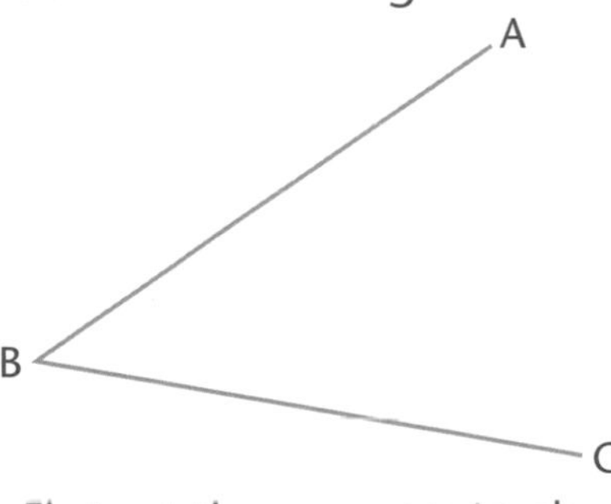

First set the compass to about 3 cm.
From B, draw arcs on BA and BC.
Where these arcs cross BA and BC, draw two further arcs to cross each other.
Draw a line from B through the point where these arcs cross.
This is the angle bisector of ABC.

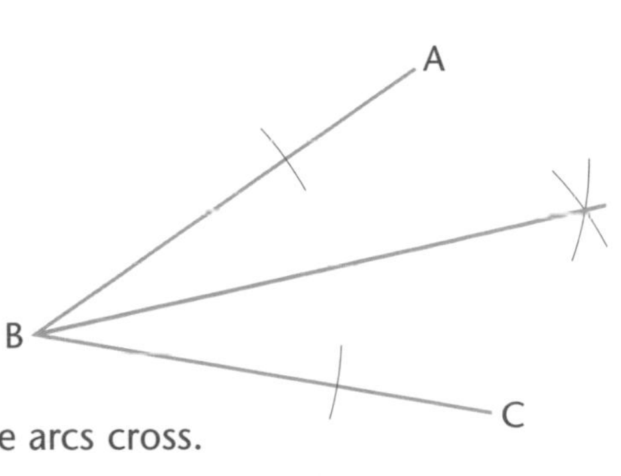

Constructing an angle of 60°

level 7

Example: Draw an angle of 60° at the point B.

Set the compass to about 3 cm.
Draw an arc from B that crosses the line and draws almost a quarter circle.
From the point where the arc crosses the line, draw another arc to cross the first.
Join the point where these arcs cross to B.
The angle at B is 60°.

B

Did You Know?

About 4000 years ago, the Babylonians tracked the path of the Sun across the sky and realised it took a year (about 360 days) to complete one circuit. This led them to divide the circle into 360°.

Sample worked test question

level 7

Construct a triangle that has the following properties:

- Total length of three sides is 12 cm.
- Only two of the sides are equal length.
- All sides are whole numbers of centimetres.

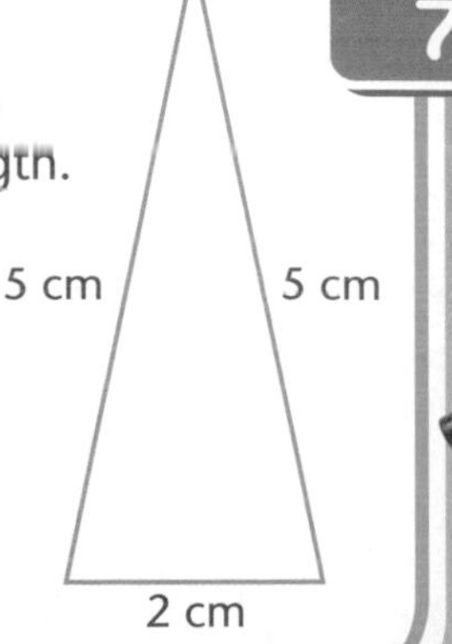

Answer
There is only one possible answer.
A triangle with sides of 2 cm, 5 cm and 5 cm.

Spot Check

1 Draw this triangle accurately.

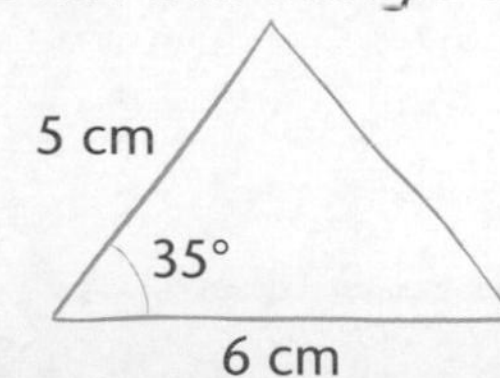

Loci

level 7

Paths

- A **locus** (singular of loci) is the **path** moved by a point according to a **rule**.

Example: Draw the locus of all the points that are

a exactly 2 cm from A

b within 3 cm of B.

A• •B

a Points that are exactly 2 cm from A form a circle of radius 2 cm centred on A.

b Points that are within 3 cm of B are all points inside a circle of radius 3 cm centred on B.

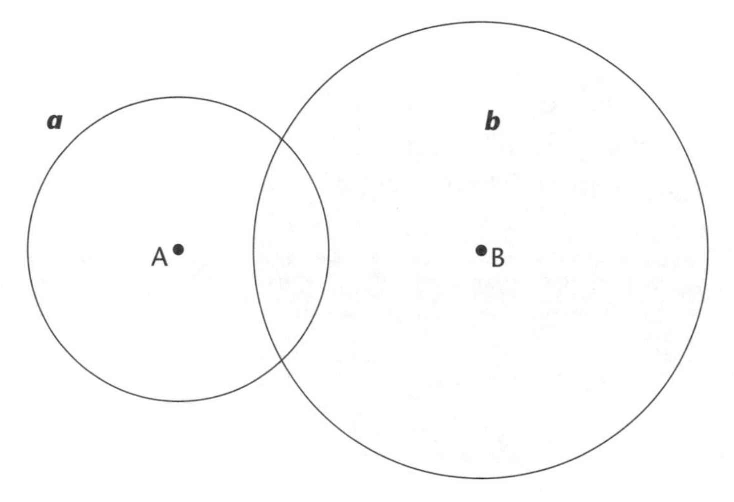

Example: Draw the locus of all points that are

a exactly 2 cm from the line AB

b the same distance from A as from B.

A•- - - - - - - - - - -•B

a The points that are exactly 2 cm from AB form a 'sausage' shape around AB with two straight lines 2 cm away each side and two semi-circles of radius 2 cm centred on A and B.

b The points that are the same distance from A and B are the points on the perpendicular bisector of AB.

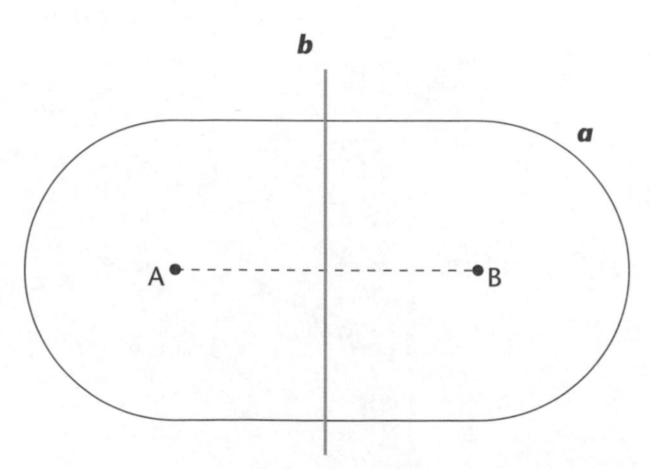

level 8

Loci

- Most loci problems are set in a real-life context.

Example: A radio transmitter is to be built so that it is the same distance from two towns: Radville and Seeton. It also has to be within 20 km of a third town, Towton. Show the possible location of the transmitter.

'Same distance from' means 'the perpendicular bisector of'.

'Within 20 km' means inside a circle of radius 20 km.

The 'overlap' of these two conditions is shown with the red line.

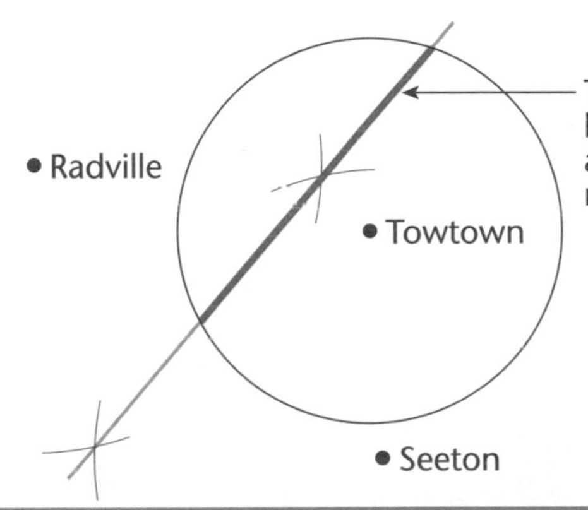

Transmitter could be positioned anywhere on this red line

Scale: 1 cm represents 10 km

Top Tip!

Make sure your construction arcs are shown and that the required locus is clearly marked.

level 8

Sample worked test question

The plan shows a garden. Each square is 1 m by 1 m.
There are four trees in the garden whose trunks are marked by T.
John wants to erect an aerial for his short wave radio.
The aerial cannot be

- within 2 metres of any tree trunk
- nearer than 1 metre to the edge of the garden.

Show the places where the aerial could be placed.

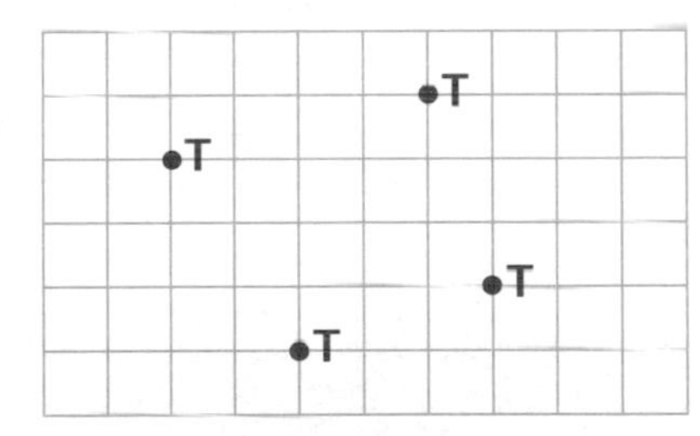

Answer
A circle of radius 2 m must be drawn round each tree and all the area within 1 metre of the edge must be excluded. The prohibited areas are shaded.

The area that is unshaded is where the aerial could be erected.

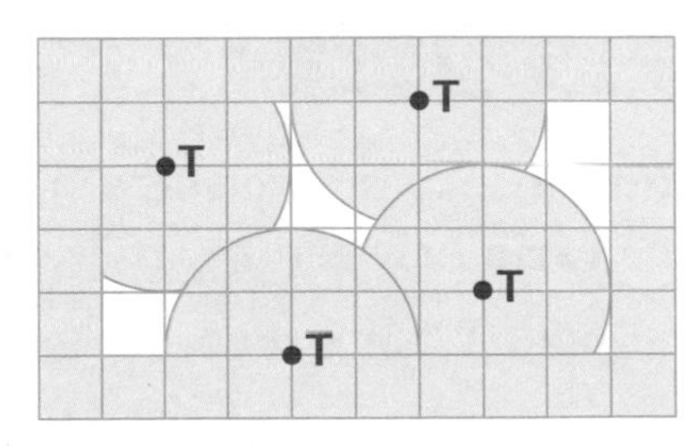

Did You Know?

The locus of the Earth around the Sun is not circular but elliptical.

1 A remote-controlled car moves on a floor so that it is always the same distance from a fixed point. What shape does the car make?

SHAPE, SPACE AND MEASURES

Similarity

Congruency and similarity

level 7

- Two shapes are **congruent** if they are **exactly** the **same size**. They may be reflected or rotated but all the sides and angles must be the same.
- Two shapes are **similar** if the sides are in the **same ratio** and the **angles** are the **same**.

Example: Which of the following triangles are **a** congruent **b** similar. Give reasons for your choices. All lengths are centimetres.

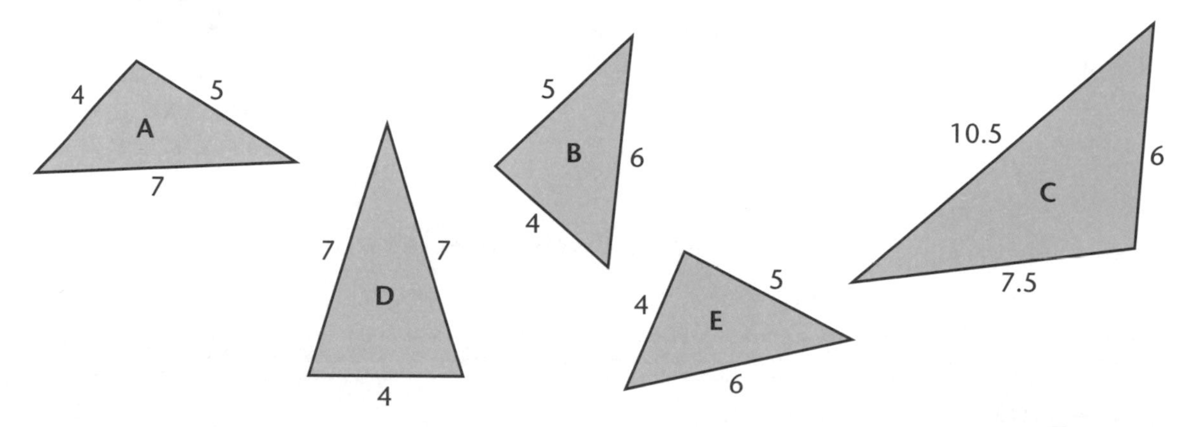

a Triangles B and E are congruent because the sides are the same.

b Triangles A and C are similar because the sides are in the same ratio.

Top Tip!

Check that **all** sides are in the same ratio and not just one or two.

Scale factors

level 7

- **Similar** shapes have **sides** in the **same ratio**.
- The **ratio** between any **pair of equivalent sides** is known as the **scale factor**.

Example: What is the scale factor between triangles A and C above?

Each side of triangle C is one and a half times the size of the sides in triangle A.
So, the scale factor is 1.5.

1 In the diagram AB is parallel to DE.
CB = 4 cm, AB = 5 cm, DE = 10 cm, CD = 9 cm.

a Explain why ABC and DEC are similar triangles.

b Find the length CE.

c Find the length AD.

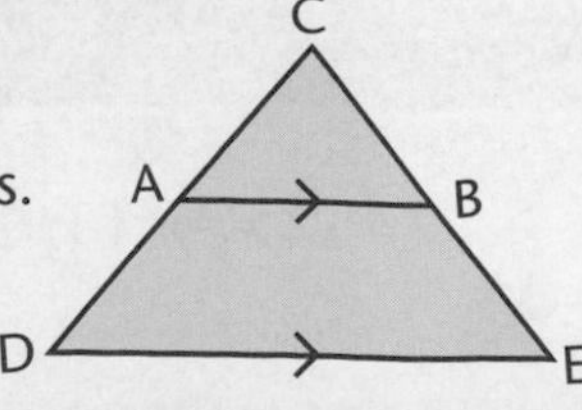

Similar triangles

level 8

- When two triangles are similar and the scale factor is known, this can be used to work out missing sides.

Example: Triangles ABC and PQR are similar.

a What is the scale factor between the lengths of the two triangles?

b Find the values of

i angle x **ii** side y **iii** side z.

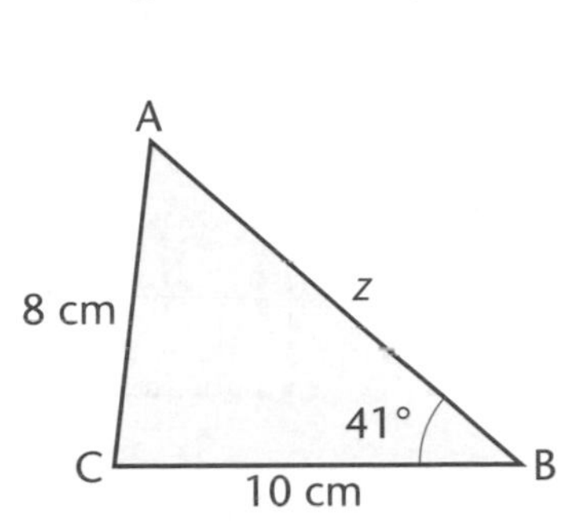

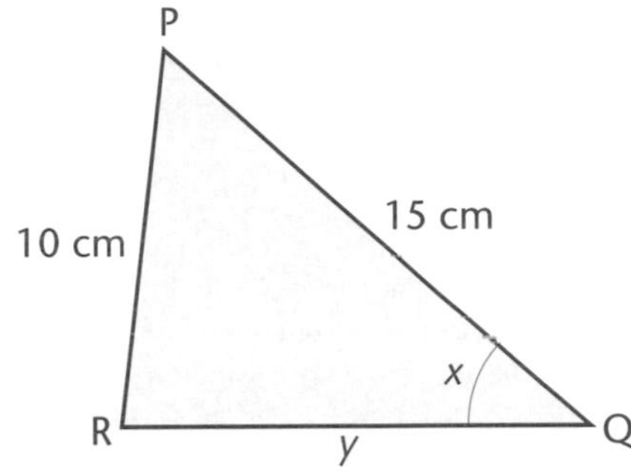

a Because AC and PR are equivalent sides, 8 x scale factor = 10.
So, the scale factor is 1.25.

b ***i*** Angles do not change so angle x is 41°.

ii Side y is 1.25 x 10 = 12.5 cm.

iii Side z is 15 ÷ 1.25 = 12 cm.

Top Tip!

The lettering will usually give you a clue. 'ABC similar to PQR' tells you that AB and PQ are equivalent sides.

Sample worked test question

level 8

Find the length AC.

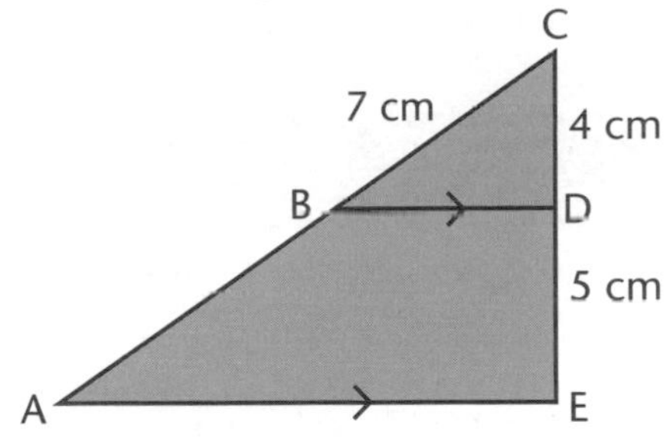

Answer

Let AC be the length x.

If we split the triangle into two separate triangles we get

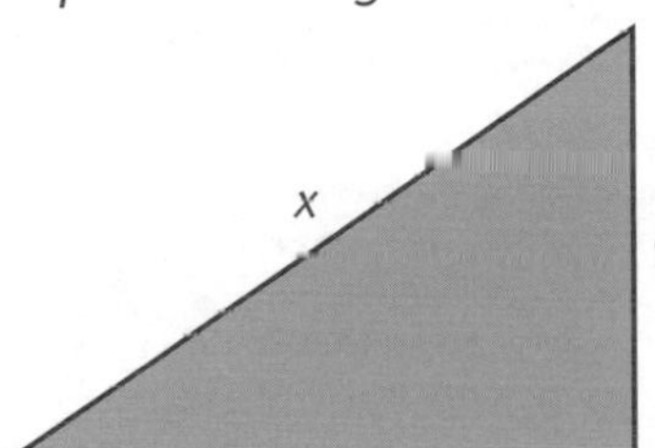

and

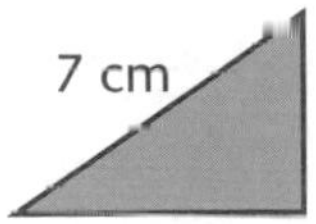

From these we can set up the equation $\frac{x}{7} = \frac{9}{4}$.

This can be rearranged as $x = \frac{63}{4}$.

This gives $x = 15.75$ *cm.*

Did You Know?

A car built in the 1960s produced about 20 times the amount of pollution that a car built today does.

Pythagoras' theorem

level 7

Pythagoras' theorem

- Pythagoras' theorem connects the **lengths** of the sides of a **right-angled triangle**.
- The theorem states:

  The area of a square on the long side of a right-angled triangle is equal to the area of the squares on the two short sides.

  Area C = Area A + Area B

  This is usually written as:

  $c^2 = a^2 + b^2$

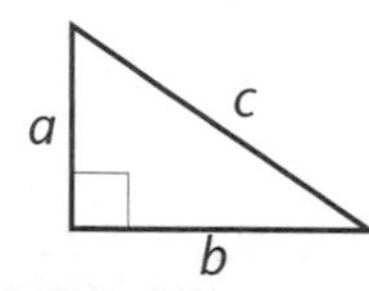

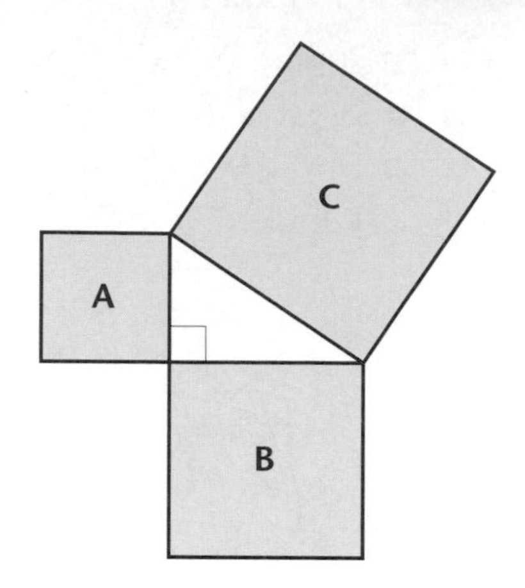

level 7

Finding the hypotenuse

- The **hypotenuse** is the **longest side** of a right-angled triangle. It is always the side opposite the right angle.

Example: Find the length x in this triangle.

5 cm, 7 cm, x

Using Pythagoras' theorem $x^2 = 7^2 + 5^2$

⇨ $x^2 = 49 + 25 = 74$

⇨ $x = \sqrt{74} = 8.60$ cm (3 s.f.)

Top Tip!

If the question doesn't state an accuracy for the answer, give it to 3 significant figures.

level 7

Finding a short side

Example: Find the length x in this triangle.
Give your answer to 1 decimal place.

4 cm, 9 cm, x

Using Pythagoras' theorem $9^2 = 4^2 + x^2$

⇨ $x^2 = 9^2 - 4^2 = 81 - 16 = 65$

⇨ $x = \sqrt{65} = 8.1$ cm

Top Tip!

When finding a hypotenuse, add the squares of the sides. When finding a short side, subtract the squares of the sides.

level 7

Pythagorean triples

- Some sets of whole numbers work for Pythagoras' theorem.
- The most famous of these is 3, 4, 5 because $5^2 = 3^2 + 4^2$.
- Others are 5, 12, 13 and 7, 24, 25 and multiples such as 6, 8, 10.

Top Tip!

It is useful to recognise the Pythagorean triples but if you can't you can always work it out.

Sample mental test question

Write down the value of $\sqrt{4^2 + 3^2}$.

$4^2 + 3^2 = 16 + 9 = 25$ and $\sqrt{25} = 5$

level 7

Sample worked test question

Is it possible for these two triangles to fit together exactly along the edge shown?

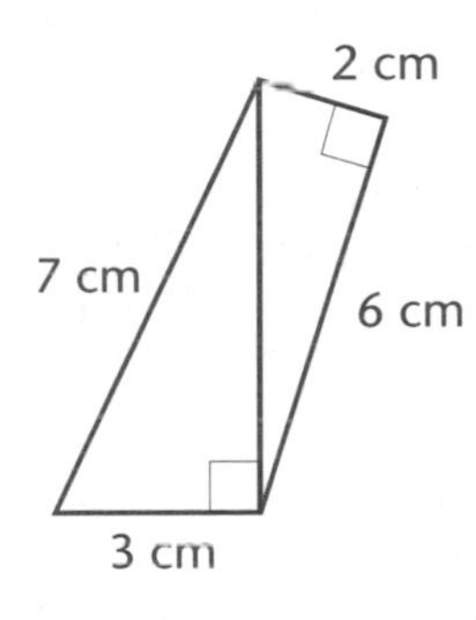

Answer

Let the common side be length x.

Looking at the left-hand triangle $x^2 = 7^2 - 3^2 = 49 - 9 - 40$, so $x = \sqrt{40}$.

Looking at the right-hand triangle $x^2 = 6^2 + 2^2 = 36 + 4 = 40$, so $x = \sqrt{40}$.

Therefore the two triangles can fit together as shown.

Did You Know?

The surveyors who built the pyramids used a knotted rope that they could make into a 3, 4, 5 triangle to set up right angles. 'Surveyors' means 'rope stretchers' in ancient Egyptian.

1 Work out the missing length x in these triangles.
Give your answers to 3 significant figures.

a

6 cm

x

9 cm

b

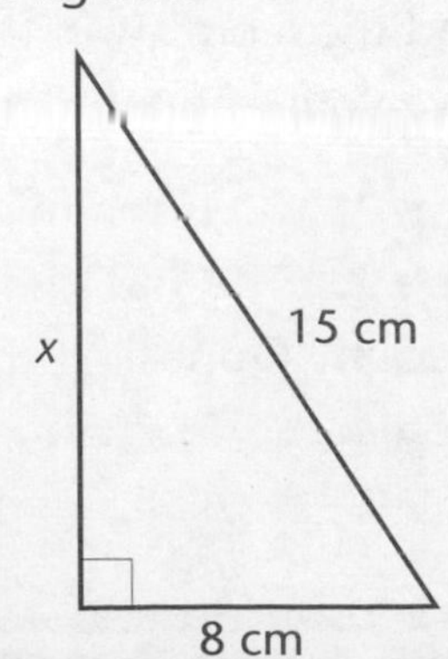

Trigonometry

level 8

Trigonometry

- Trigonometry connects the **sides** and the **angles** of a right-angled triangle.
- In this triangle:
  side a is **opposite** (opp) the angle
  side b is **adjacent** (adj) to the angle
  side c is the **hypotenuse** (hyp).

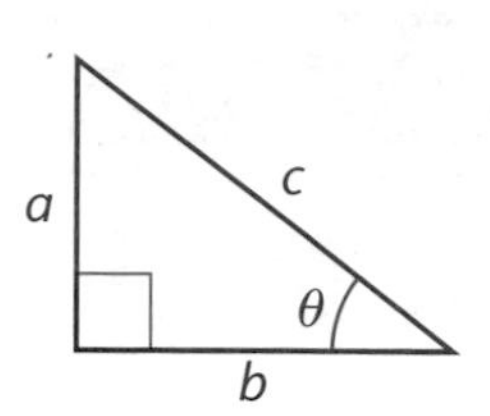

- There are three trigonometric ratios, **sine** (sin), **cosine** (cos) and **tangent** (tan), which are connected by the following rules:

$\sin\theta = \frac{\text{opp}}{\text{hyp}} = \frac{a}{c}$ $\cos\theta = \frac{\text{adj}}{\text{hyp}} = \frac{b}{c}$ $\tan\theta = \frac{\text{opp}}{\text{adj}} = \frac{a}{b}$

Each ratio has three possible relationships. For sine these are:

$\sin\theta = \frac{\text{opp}}{\text{hyp}}$ opp = hyp x sin $\text{hyp} = \frac{\text{opp}}{\sin\theta}$

Top Tip!

You will need to remember these. Most people use the mnemonic SOH CAH TOA, pronounced SOH-CA-TOE-AH.

level 8

Solving trigonometric problems

- To solve a trigonometric (trig) problem there are four steps.
  Step 1: Relative to the angle, identify the names of the sides in the problem i.e. which two of opposite, adjacent or hypotenuse are involved.
  Step 2: Identify which trig ratio links these two sides i.e. sine, cosine or tangent. There is only one ratio for every possible pair of sides.
  Step 3: Set up the calculation.
  Step 4: Work it out.

level 8

Finding a side

Example: Find the length x in this triangle.
Give your answer to 1 decimal place.

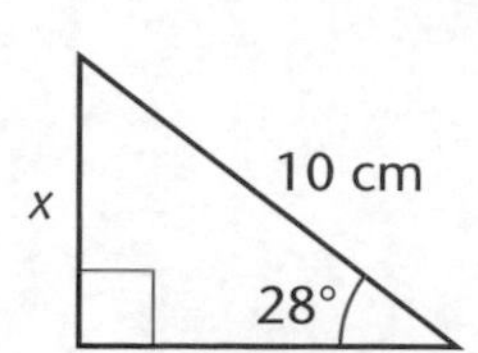

Step 1: 10 is the hypotenuse and x is the opposite.
Step 2: This is a sine problem.
Step 3: Opposite = hypotenuse x sin θ, so $x = 10 \times \sin 28$.
Step 4: $x = 4.7$ cm

Top Tip!

If the question doesn't state an accuracy for the answer, give it to 3 significant figures.

level 8

Finding an angle

Example: Find the angle x in this triangle.

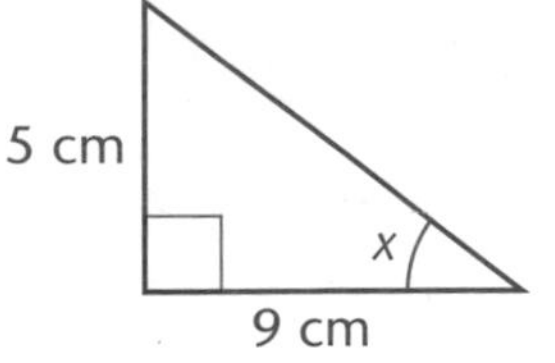

Step 1: 5 is the opposite and 9 is the adjacent.
Step 2: This is a tangent problem.
Step 3: $\tan x$ = opposite ÷ adjacent, so $\tan x = 5 \div 9 = 0.5555$.
Step 4: $\tan x = 0.5555$
➪ $x = \tan^{-1} 0.5555$
➪ $x = 29.1°$

$x = \tan^{-1} 0.5555$ means the inverse tan of 0.5555, i.e. the angle that has a tan of 0.5555.

level 8

Sample worked test question

Two towns are shown on the map.
Manton is 8.2 km east and 7.7 km north of Norville.
What is the bearing of Manton from Norville?

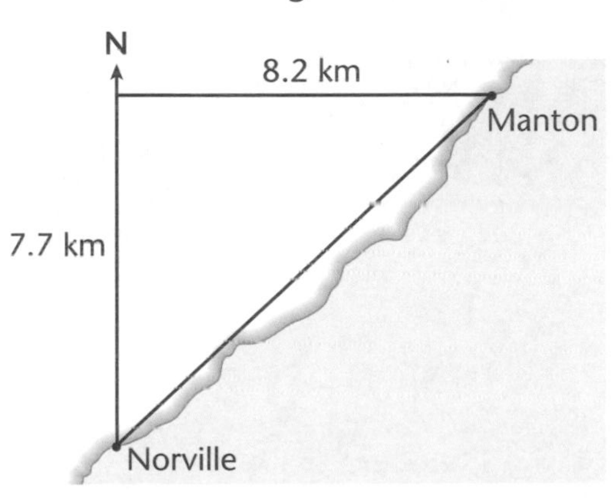

Answer
The angle is the clockwise angle measured from north.
Using the steps of the trig process:
Step 1: *7.7 is the adjacent and 8.2 is the opposite.*
Step 2: *This is a tangent problem.*
Step 3: *$\tan x$ = opposite ÷ adjacent, so $\tan x = 8.2 \div 7.7 = 1.0649$.*
Step 4: *$\tan x = 1.0649$*
➪ *$x = \tan^{-1} 1.0649$*
➪ *$x = 46.8°$*

Did You Know?

Trigonometry was created by the ancient Greeks in about 190–120 BC so they could study the position of the planets.

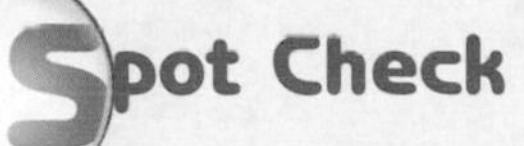

1 Find the side x in this triangle.

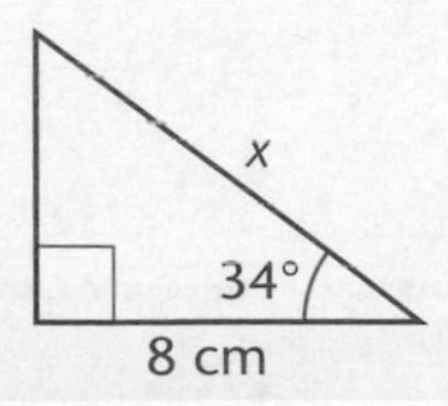

SHAPE, SPACE AND MEASURES

Sectors and circle theorems

level 8

Sectors

- A **sector** of a circle is formed by **two radii** from the centre.
- If the angle between the two radii is θ, the **length** and **area of arc** are given:

Length of arc = $\frac{\theta}{360} \times 2\pi r$

Area of sector = $\frac{\theta}{360} \times \pi r^2$

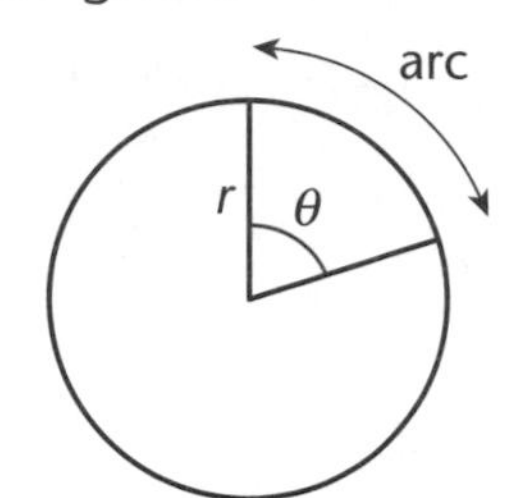

Example: Find **a** the perimeter and **b** the area of a sector of a circle with a radius of 10 cm and an angle of 72°.

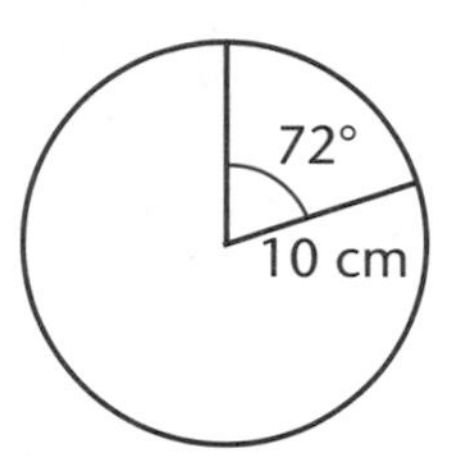

a Length of arc = $\frac{72}{360} \times 2 \times \pi \times 10 = 12.6$ cm.
Perimeter of sector = 12.6 + 10 + 10 = 32.6 cm.

b Area = $\frac{72}{360} \times \pi \times 10^2 = 62.8$ cm^2

Top Tip!

The perimeter is the **total length** around a shape so you need to add the two radii as well.

level 8

Circle theorems

- There are four circle theorems you need to know.

Theorem 1

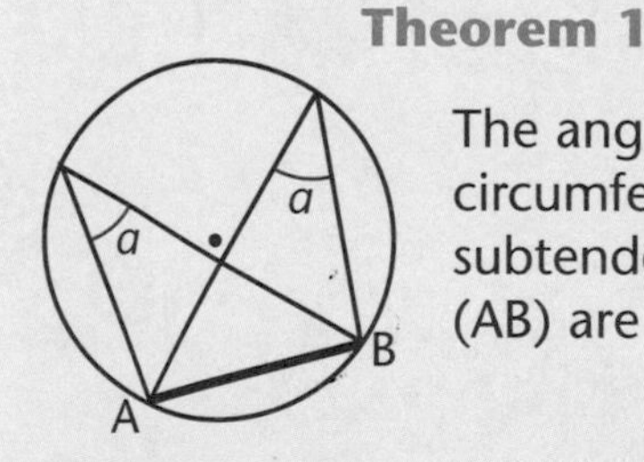

The angles at the circumference subtended by a chord (AB) are equal.

Theorem 2

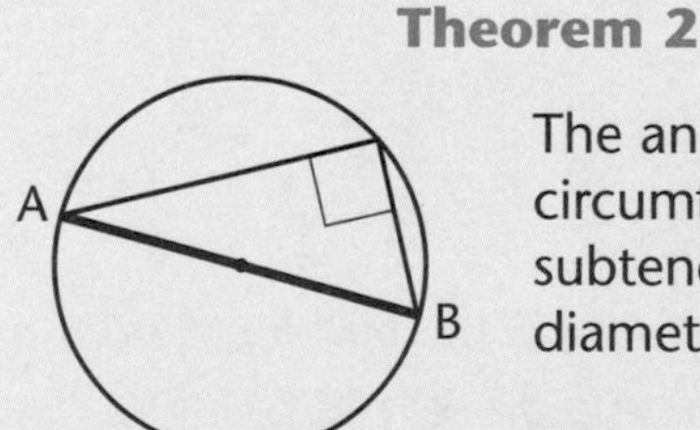

The angle at the circumference subtended by a diameter (AB) is 90°.

Theorem 3

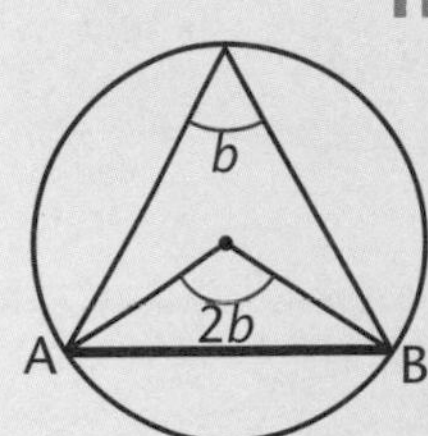

The angle subtended at the centre by a chord (AB) is twice the angle at the circumference.

Theorem 4

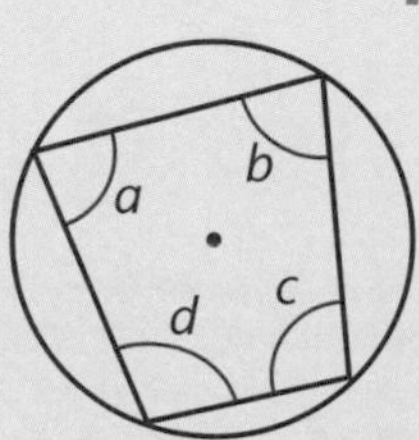

Opposite angles in a cyclic quadrilateral add to 180°.
$a + c = b + d = 180°$

Circle theorems continued

Example: Find the angles marked in the following circles. Give a reason for your answer.

a

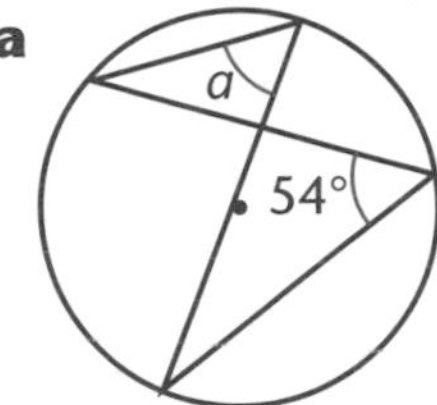

b

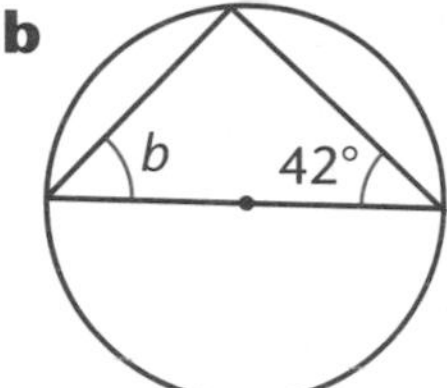

c

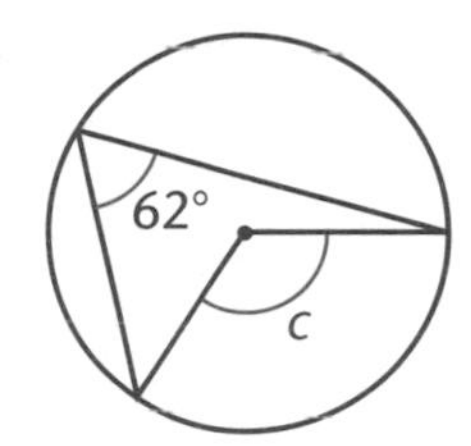

d

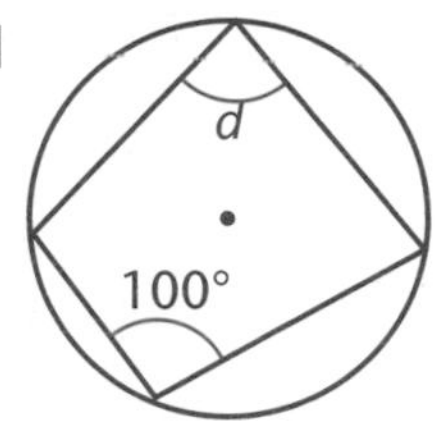

a Angle $a = 54°$ because it is an angle on the same chord as the angle of 54° shown in the diagram (Theorem 1).

b Angle $b = 48°$ because the angle at the circumference is 90° (Theorem 2) and there are 180° in a triangle.

c Angle $c = 124°$ because it is twice the angle at the circumference (Theorem 3).

d Angle $d = 80°$ because it is an angle in a cyclic quadrilateral opposite an angle of 100° (Theorem 4).

level **8**

Sample worked test question

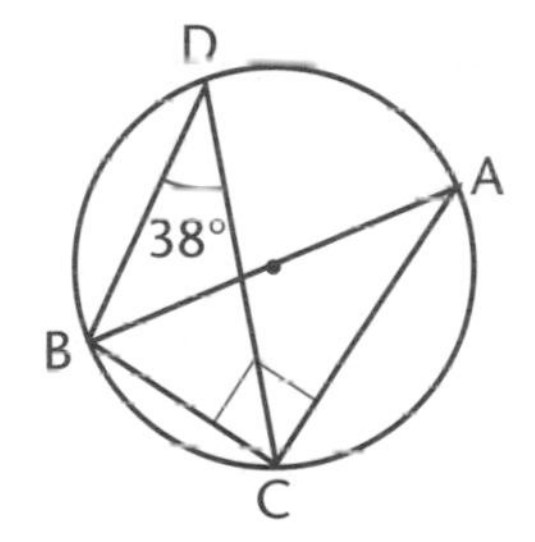

ABC is a right-angled triangle that fits within a circle.

D is a point on the circumference.

Angle CDB is 38°.

a Explain why AB must be a diameter.

b Work out the size of angle CBA.

Answers

a *Angle C is a right angle so AB must lie on a diameter.*

b *Angle BAC = 38° (angles on same chord).*
So, angle CBA = 52° (angles in a triangle).

Did You Know?

Crop circles first appeared in the 1980s and nobody has yet been able to explain what causes them. Look them up on the Internet.

Spot Check

1 AB is a chord of a circle radius 6 cm with an angle of 100° at the centre.

a Find the length of the arc AB.

b Find the angle x. Give a reason for your answer.

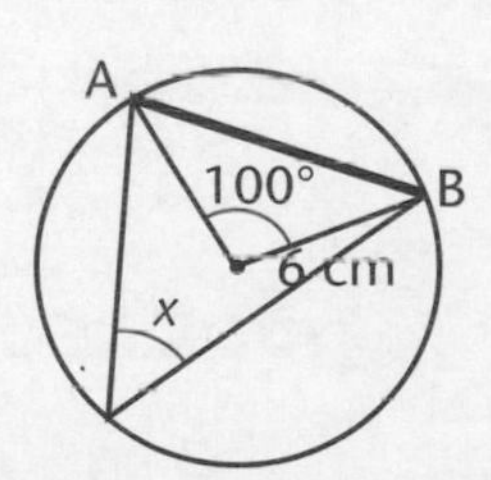

Comparing distributions

levels 5-6

The mean

- To find the **mean** for a set of data, first find the **total** of all the **values** and then **divide** this total by the **number of values**.

The symbol for the mean is $\bar{x}$. $\bar{x} = \frac{\text{total of all values}}{\text{number of values}}$

Example: The ages of six people are: 21, 32, 25, 19, 23 and 18. Find their mean age.

$\bar{x} = \frac{\text{total of all values}}{\text{number of values}} = \frac{138}{6} = 23$

level 6

Comparing distributions

- You compare distributions in everyday situations, without even realising it.

Example: Two dinner ladies, Mary and Doris, serve chips in the school canteen.

Rajid went to Mary for his chips for a week. Mary gave out 18, 23, 25, 25, 34 chips.

The following week Rajid went to Doris for his chips. Doris gave out 23, 25, 27, 25, 25 chips.

Which dinner lady should Rajid go to to be given the most chips?

First, look at the averages and range:

	Mean	Median	Mode	Range
Mary	25	25	25	16
Doris	25	25	25	4

The averages are all the same but Mary's range is much larger than Doris'. So if Rajid caught Mary on a good day, he might have as many as 34 chips, but on a bad day he might have as few as 18. Doris is very consistent and will always give about 25 chips.

You could say, 'The averages are the same' and 'I would go to Mary because she has a bigger range and you might be lucky and get a lot of chips', or you could say, 'I would go to Doris because she has a smaller range and is more consistent'. It doesn't matter who you choose as long as you mention the **averages** and the **range** and give **reasons** for your choice.

Top Tip!

The **range** measures the spread of the data so gives an indication of how **consistent** the data is.

Top Tip!

When comparing data using ranges and averages, you must **refer** to them both in your answer.

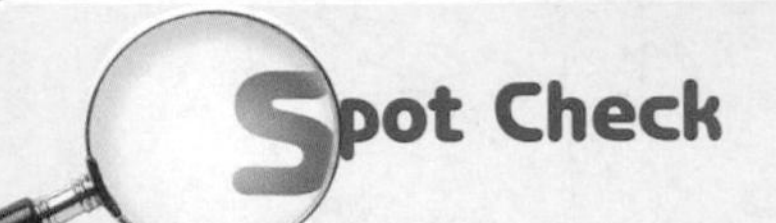

1 Work out the mean and range of these two sets of data.

a 4, 6, 6, 9, 10

b 2, 5, 7, 8, 13

levels 5-6

The mean and median from a frequency table

- To find the mean from a frequency table, add an extra column to find the total of all the values.

Example: The frequency table shows the marks for 20 students in a spelling test. Find the mean mark and the median mark.

Mark, x	Frequency, f	$x \times f$
5	1	5
6	0	0
7	3	21
8	5	40
9	8	72
10	3	30
Totals	20	168

Example: $\bar{x} = \frac{\text{total of all values}}{\text{number of values}} = \frac{168}{20} = 8.4$ (mean)

The median is between the 10th and 11th values. Counting up the frequency column gives 10th and 11th values as 9. Median = 9.

Top Tip!

If the mean is not an exact answer, then round it to 1 decimal place.

Sample mental test question

Give three numbers with a mode of 5 and a range of 2.

If there are three numbers and 5 is the mode, then two of the numbers must be 5.
To give a range of 2, the other number must be 3 or 7. So there are two answers: 3, 5, 5 or 5, 5, 7.

Sample worked test question

Jayne needs to pick an attacker for the netball team.
She looks at the scoring record of Asha and Rhoda.
In Asha's last five matches she scored 5, 7, 2, 9, 2 goals.
In Rhoda's last five matches she scored 5, 6, 5, 4, 5 goals.
Who should Jayne choose and why?

Answer
The averages for Asha are: Mode 2, Median 5, Mean 5.
The averages for Rhoda are: Mode 5, Median 5, Mean 5.
The range of Asha's scores is 9 – 2 = 7.
The range of Rhoda's scores is 6 – 4 = 2.
The averages are the same, except for the modes, but Rhoda has a smaller range so she is more consistent.
Although Asha may get a high score, she may also get a low score so Jayne should pick Rhoda.

Did You Know?

On an average work day, a typist's fingers travel 20 kilometres, over a range of 20 cm.

Line graphs

level 5

Plotting values

- A **line graph** is a clear way of showing **changes in data.**

Example: The maximum temperature in a town each month for a year is recorded.

Jan	Feb	Mar	Apr	May	Jun	Jul	Aug	Sep	Oct	Nov	Dec
4	9	13	19	25	28	32	29	22	17	13	7

a Which two months had a maximum temperature of 13 °C?

b Which was the hottest month?

c What is the difference between the hottest and coldest months?

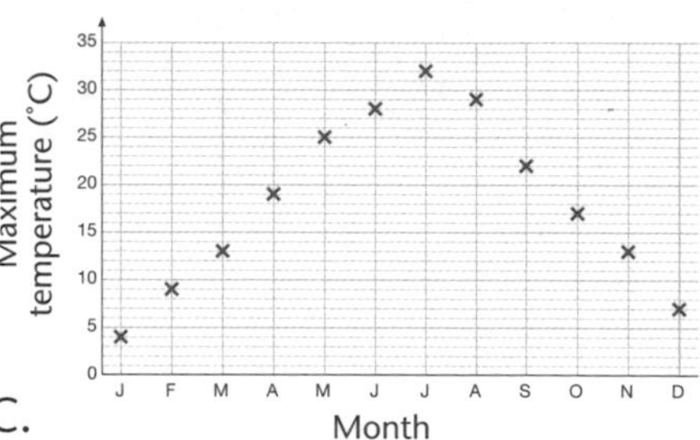

a Reading across from 13 °C on the y-axis gives the two months as March and November.

b July was the hottest month with a maximum temperature of 32 °C.

c January was the coldest month with a maximum temperature of 4 °C so the difference is 32 – 4 = 28 °C.

level 5

Trend lines

- You can see from the graph above that the temperature rises in summer. If we join the points, the lines between them have no meaning but they show the **trend** of the temperatures over the year.

Example: The graph below shows Sam's height from the age of 2 to the age of 8.

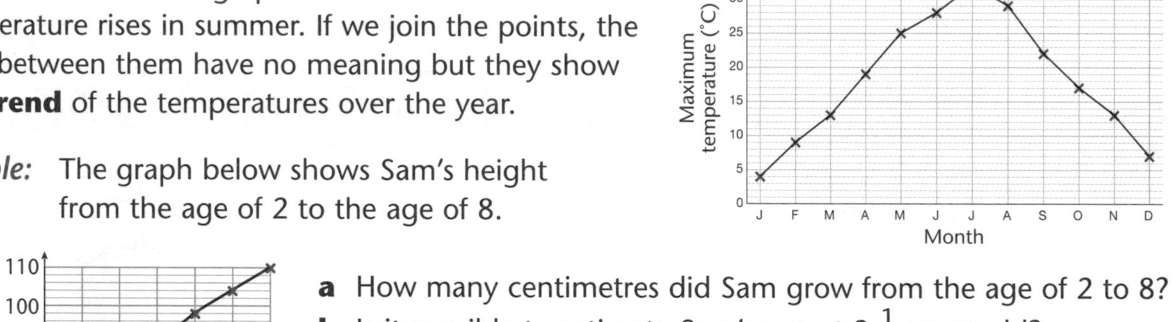

a How many centimetres did Sam grow from the age of 2 to 8?

b Is it possible to estimate Sam's age at $3\frac{1}{2}$ years old?

a Sam was 54 cm at age 2 and 110 cm at age 8 so he grew 56 cm.

b Yes, growth is continuous so the line has a meaning this time. Sam was about 74 cm tall at age $3\frac{1}{2}$.

Example: The following graph shows the monthly gas use for the Henman family.

a During which month was the most gas used?

b During one of the months in the summer the Henman family went on holiday. Which month was this? Give a reason for your answer.

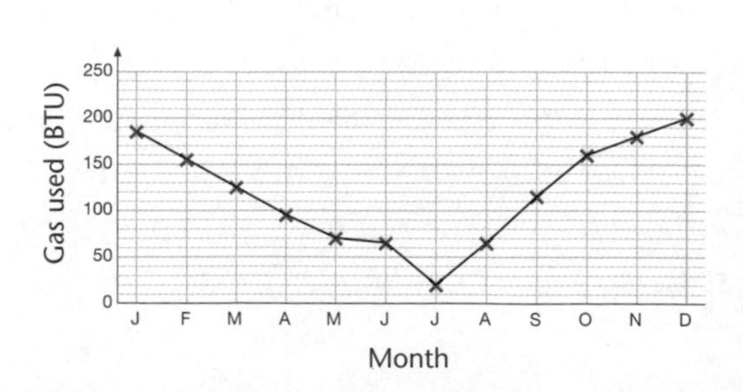

a The most gas was used during December as this is the highest value on the graph.

b July. The amount of gas drops dramatically in July suggesting that the family were not at home.

level 5

Sample worked test question

Jason records the temperature in his greenhouse once an hour. At 8 am it was 14 °C, at 9 am it was 20 °C, at 10 am it was 25 °C and at 11 am it was 29 °C.

This information is shown on the graph.

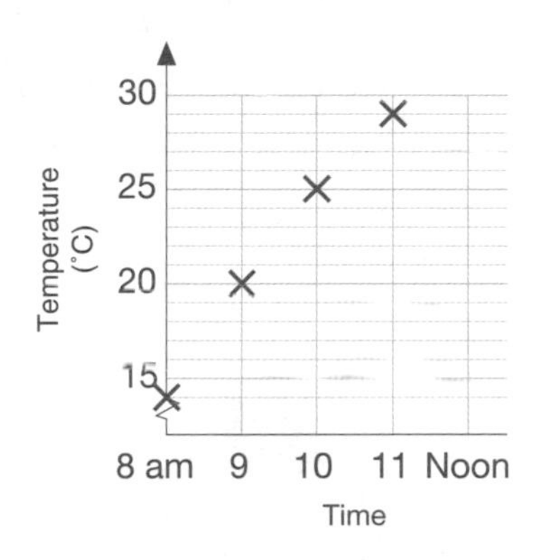

a Estimate the temperature at 10.30 am.

b Explain why the graph cannot be used to predict the temperature at 12 noon.

Answers

a *Using the trend line between 10 am and 11 am, the temperature can be estimated as 27 °C.*

b *The trend line may not continue after 11 am. The sun could go in, or the windows could be opened.*

Top Tip!

You can **estimate** values from trend lines but you cannot say for sure what the values are.

Did You Know?

The world's tallest person grew to a height of 270 cm.

Spot Check

1 The temperature drops by 4 °C from 5 am to 6 am. It was 7 °C at 5 am.

a What was the approximate temperature at 5.30 am?

b Can you estimate the temperature at 7 am?

2 Using the graph of Sam's height on the opposite page, decide during which years Sam grew the fastest? Explain how you can tell.

HANDLING DATA Statistical diagrams

level 5

Drawing pie charts

Example: This table shows the types of vehicles parked in a motorway service area.
Draw a pie chart to show the data.

Type of vehicle	Frequency
Car	40
Vans	22
Motorbikes	8
Lorries	20

First add up the frequencies: they total 90.

Divide this into 360 to find the angle that represents each vehicle: $360 \div 90 = 4°$.

Now multiply each frequency by this figure. This is easily shown by adding another column to the table.

Type of vehicle	Frequency	Angle
Car	40	40 x 4 = 160°
Vans	22	22 x 4 = 88°
Motorbikes	8	8 x 4 = 32°
Lorries	20	20 x 4 = 80°

Top Tip!
Check the angles add to 360.

Then start with a circle, draw a radius and measure each angle in turn.

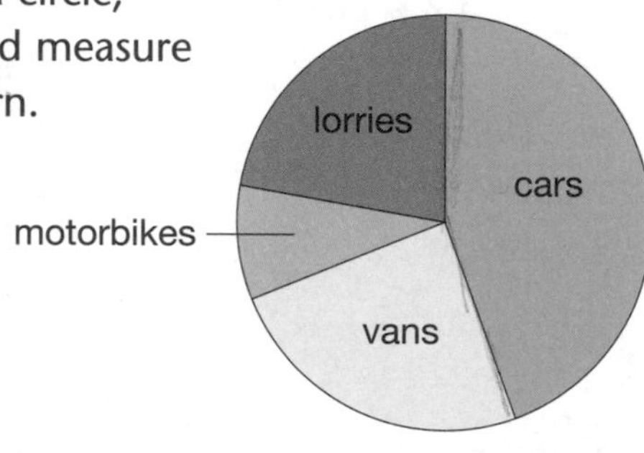

Top Tip!
It is useful to know the factors of 360:

1 x 360	8 x 45
2 x 180	9 x 40
3 x 120	10 x 36
4 x 90	12 x 30
5 x 72	15 x 24
6 x 60	18 x 20

level 6

Stem-and-leaf diagrams

- A **stem-and-leaf diagram** shows **ordered** data in a **concise** way.
- The **stem** is the **10s digit** and the **leaves** are the **units**.

Example: Show the data 32, 41, 56, 37, 38, 29, 42, 46, 38, 28, 34, 38, 37, 51, 49 on a stem-and-leaf diagram.

Stem	Leaves						
2	8	9					
3	2	4	7	7	8	8	8
4	1	2	6	9			
5	1	6					

Key: 2 | 8 represents 28

Top Tip!
Always put a **key** on a stem-and-leaf diagram.

Sample mental test question

The pie chart shows the ratio of men to women at a concert.

If 2000 people attended, how many women were there?

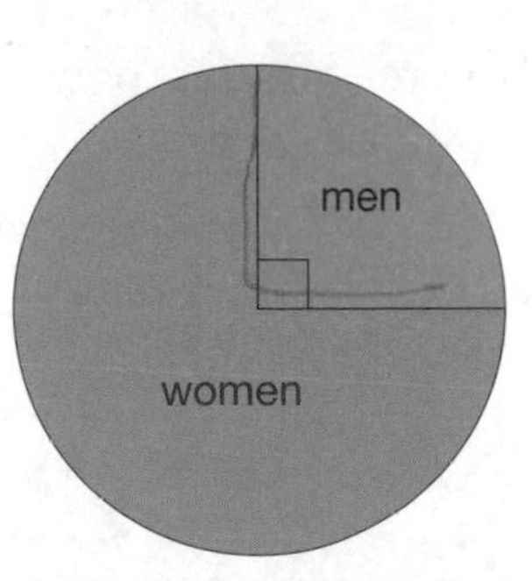

You can see from the pie chart that 75% of the people were women, 75% of 2000 = 1500.

level 5

Sample worked test question

The pie chart shows the results of a survey about where families went for their holidays.

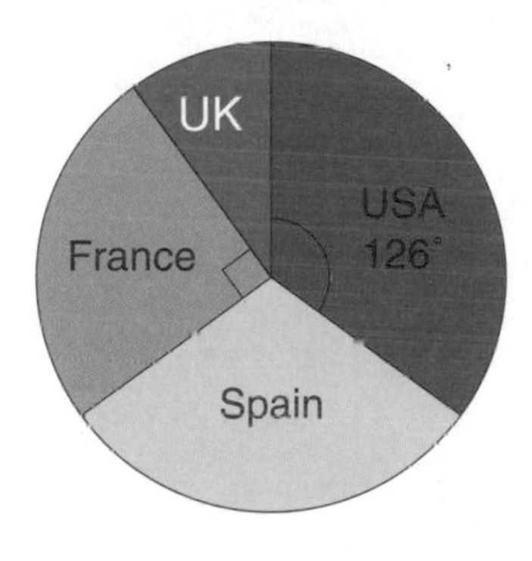

The sector for the USA represents 7 families.

How many families went to France?

Answer

There are 126° representing the USA sector.

7 families = 126°, so 1 family = 126 ÷ 7 = 18°

There are 360 ÷ 18 = 20 families in the survey.

France is a quarter of the pie chart,

so 20 ÷ 4 = 5 families went to France.

Did You Know?

The world's biggest meat pies are made in Denby Dale, Yorkshire. The last was the Millennium Pie which was claimed to weigh 12 tonnes. However, nobody will admit to have seen it and the *Guinness Book of Records* doesn't recognise it.

1 If 7 out of 30 people prefer coffee, what angle would coffee have on a pie chart showing people's favourite drinks?

2 The 'dog' sector of a pie chart representing the favourite pets of a class has an angle of 20°.

a Explain why there cannot be 30 students in the class.

b How many students could there be in the class?

Scatter diagrams

Scatter diagrams

- A **scatter diagram** shows the relationship between two variables, for example, the temperature and the sales of ice cream.
- The mathematical name for the relationship is **correlation**.
- The following diagrams show the different types of correlation.

Strong, positive correlation

Weak, positive correlation

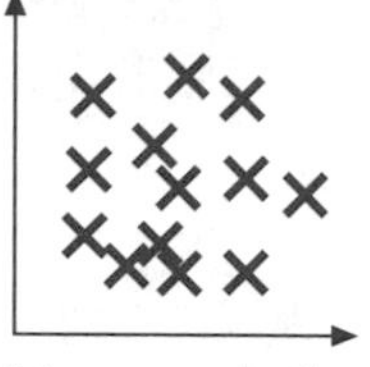

No correlation

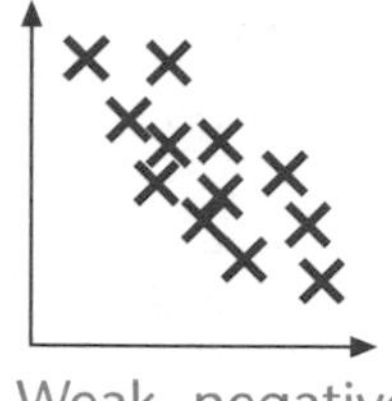

Weak, negative correlation

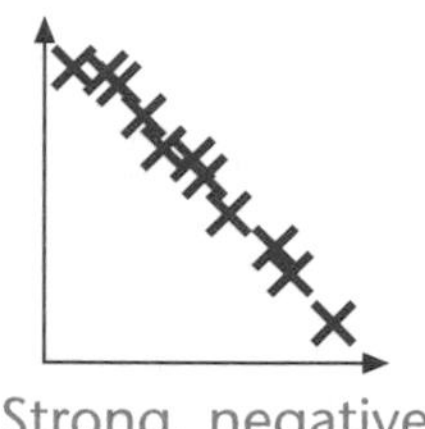

Strong, negative correlation

Example: The scatter diagram on the right shows the relationship between the cost of taxi fares and distances of journeys.

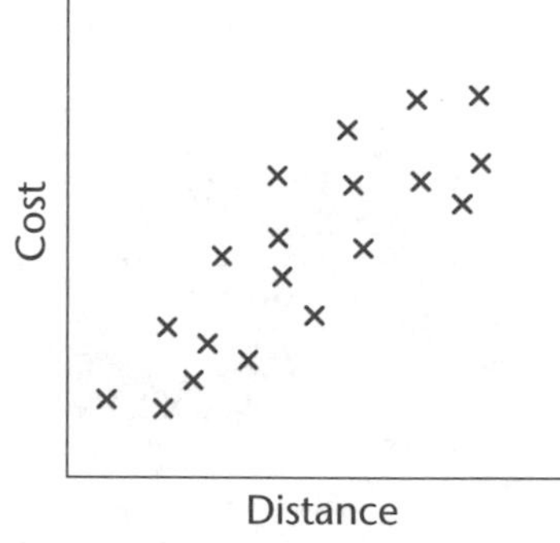

a Describe the correlation between the variables.

b Describe the expected correlation between the following:

i The cost of a taxi journey and the age of the taxi driver.

ii The time of a taxi journey and the number of cars on the road.

iii The distance of a taxi journey and the time of the journey.

a The diagram shows weak, positive correlation.

b i There will be no correlation between the age of a driver and the cost of a journey.

ii The time of a taxi journey will increase with more cars on the road, so it will show weak, positive correlation.

iii The longer a journey is the more time it will take so there is positive correlation.

Line of best fit

- A **line of best fit** is a line that passes through the data and passes as close to as many of the points as possible. It can be used to predict values.

Example: This scatter diagram shows the top speed and engine size of some cars, and a line of best fit.

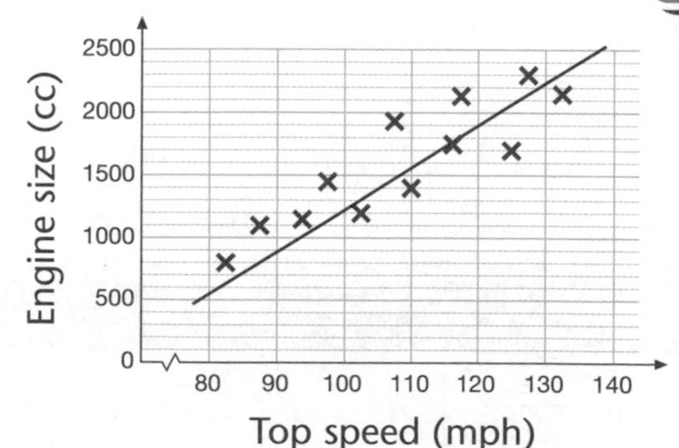

a What does the scatter diagram show about the relationship between the engine size and top speed of cars?

b The top speed of a car is 120 mph. Use the line of best fit to estimate the engine size.

a The scatter diagram shows that cars with a larger engine have a higher top speed.

b Going up from 120 mph on the Top speed axis to the line of best fit and across to the Engine size axis gives 1900 cc.

Sample mental test question

Describe the correlation in the diagram.

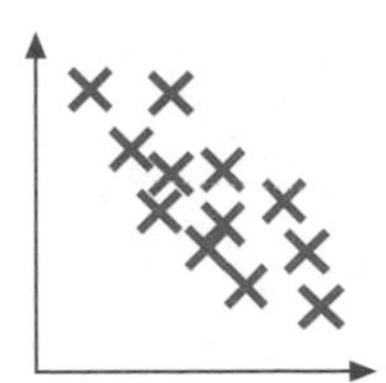

The correlation is weak negative but you would still get marks for writing 'negative correlation'.

Sample worked test question

level 6

A fish breeder keeps records of the age and mass of his prize carp. He plots the results on a scatter diagram.

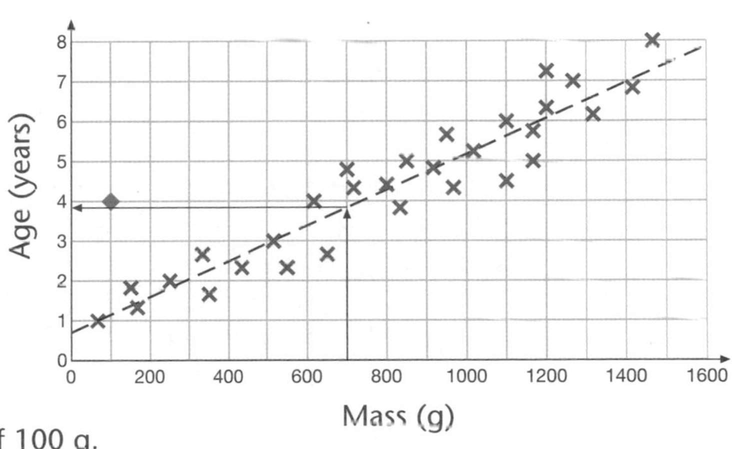

a A fish is 4 years old and has a mass of 100 g.
Explain why this fish is not likely to be a carp.

b The breeder is given a carp that has a mass of 700 g but he does not know how old it is.
He only uses fish for breeding if they are over 5 years old. Will this fish be suitable for breeding?
Give a reason for your answer.

Answers

a *The data shows strong positive correlation.*
Plot the point (100, 4). (This is shown as a diamond.)
It is clear that this point is well away from the others. It does not have the same correlation as the other values so the fish is unlikely to be a carp.

b *Draw the line of best fit. (This is shown dashed.)*
Draw a line from 700 g up to the line of best fit and then across to the Age axis. (These are the solid lines.)
This comes to just under 4 years.
Therefore the fish may not be old enough for breeding.

Did You Know?

A human can jump about four times their body length. A flea can jump 350 times its body length.

Spot Check

1 Describe the expected correlation between
a the temperature and the number of cold drinks sold
b the time of a journey and the average speed.

Surveys

level 6

Surveys

- **Surveys** are used to find out information. Groups, such as the government, need information so they can plan for the future. Companies need to know who buys their products.
- Information is usually collected using a **questionnaire**.
- For example, if you want to find out if students would like to have a party, you would want to know which day they would prefer, what type of refreshments, what type of music and how much they would pay.
- Questions in a questionnaire should be **unbiased**.

Example: This is a question on a survey about which day to hold a party.

You would prefer a party on Friday, wouldn't you?
Yes ☐ No ☐

This question is biased as it forces an opinion on the person being surveyed.

A better question would be:

On which day would you prefer a party?
Thursday ☐ Friday ☐ Saturday ☐

Top Tip!

Don't ask personal questions such as 'How old are you?' and expect an answer. People may be embarrassed to give their age.

level 6

Response sections

- Questions should have a simple response section with clear choices, no overlapping responses and a wide range of responses.

Example: Look at this question from a survey with a response section.

How old are you?
Under 10 ☐ 10–20 ☐ 20–30 ☐ Over 30 ☐

The response section of this question has overlapping catagories.

A better question would be:

How old are you?
Under 10 ☐ 11–20 ☐ 21–30 ☐ 31 or over ☐

Top Tip!

Keep questions short and with a small choice of answers. Make your responses simple so you can use tick boxes.

level 6

Sampling

- You also need to be very careful about where you undertake a survey and who you ask.

  If the school has a Friday lunchtime party and you did your survey there, you would get a **biased sample** as the students would be likely to say 'yes'.

  If you just asked a Year 7 tutor group, they might not want a party and the views of the Year 11s would not have been taken into account. This would be a **non-representative sample**.
- You should make sure the people who are surveyed are from a range of age groups and have **different views**.

Top Tip!

One way to ensure an unbiased and representative sample is to choose the people you survey **randomly**. For example, you could put all the names in a hat and draw some out. In a **random sample** everyone has an **equal chance** of being picked for the sample.

level 6

Sample worked test question

Year 9 are planning a trip and some students decide to do a survey about where people want to go.

a This is one of Ricky's questions.

Do you want to go paintballing?

Yes ☐ No ☐

What is wrong with this question?

b This is another question.

How much are you willing to spend?

£0–£10 ☐ £0–£15 ☐ Over £20 ☐

What is wrong with this question?

c Ricky decides to ask all the boys in his football practice group. What is wrong with this method of doing the survey?

Answers

a *There are not enough choices. Ricky is probably trying to get everyone to agree to go paintballing.*

b *The responses overlap so someone wanting to spend £9 would have two boxes to tick.*

c *They will all have similar opinions. The sample is non-representative and will give a biased response.*

Did You Know?

The first public opinion surveys started in 1935 and the first question ever asked was 'Do you think expenditures by the government for relief and recovery are too little, too great, or just about right?'. You could find several things wrong with this question!

Spot Check

1 Give two reasons why this is not a good question in a survey: 'Fast food makes you fat and is unhealthy. Do you agree?'

Yes ☐ No ☐

HANDLING DATA

Box plots and cumulative frequency diagrams

Box plots

level 8

- A **box plot** is a way to show the **shape** of a **distribution** and information about a set of data.
- There are five pieces of information that are shown. These are:
  - The **lowest value** of the set of data.
  - The **lower quartile** (the 'quarter' way value when in order).
  - The **median** (the 'halfway' value when in order).
  - The **upper quartile** (the 'three-quarter' way value when in order).
  - The **highest value** of the set of data.
- The **interquartile range** is the upper quartile minus the lower quartile.
- A box plot looks like this.

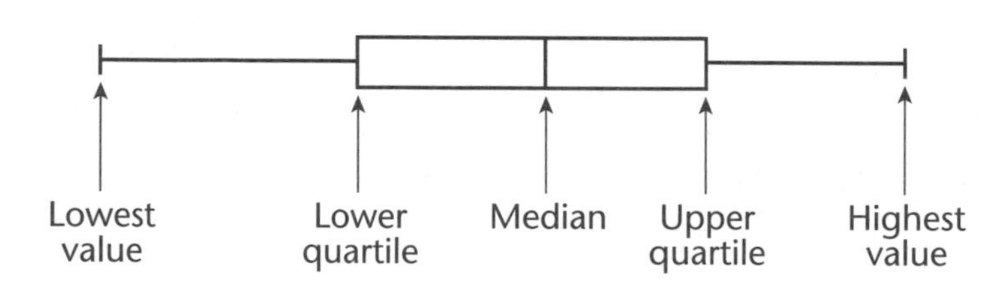

Top Tip!

A box plot is always drawn on a scale so the values can be easily read.

Example: The box plot shows the distribution of the mass of some guinea pigs.

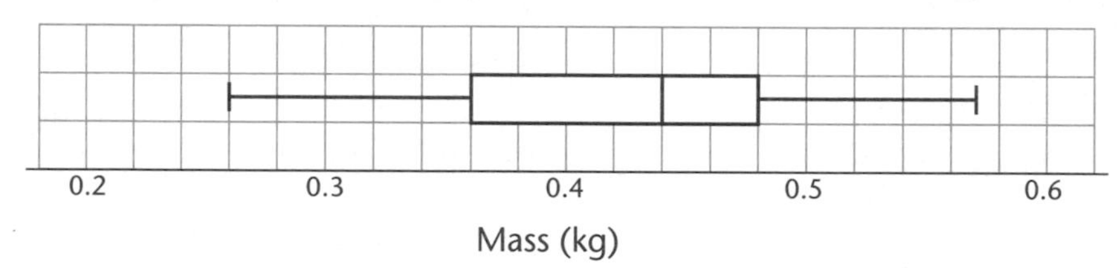

a What is the range of the data?

b What is the interquartile range of the data?

a The range is the highest value minus the lowest value.
Range = 0.57 – 0.26 = 0.31 kg.

b The interquartile range is the upper quartile minus the lower quartile. Interquartile range = 0.48 – 0.36 = 0.12 kg.

Did You Know?

The first product to have a bar code was Wrigley's chewing gum.

Spot Check

1 The same 60 students who ran the 100 metres race detailed on the opposite page did it again a week later.
The median reduced by 0.5 seconds.
The lower quartile reduced by 0.3 seconds.
The interquartile range stayed the same.
The fastest time was 11.8 seconds.
The slowest time was 18.3 seconds
Draw a box plot to show this data.

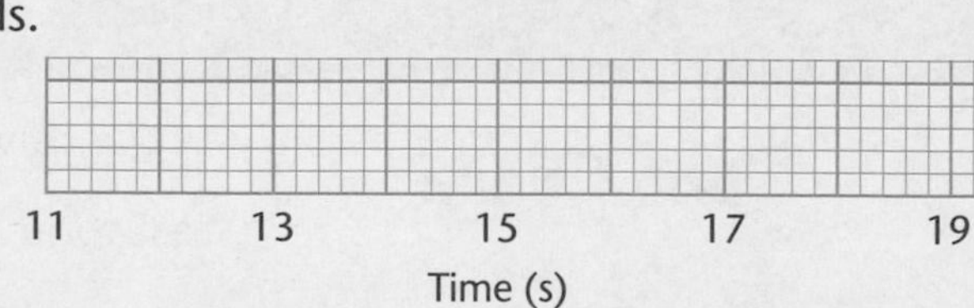

Cumulative frequency diagrams

level 8

- A **cumulative frequency diagram** is a way of representing **continuous data** so that the median, quartiles and interquartile range can be found.

Example: The table shows the times that 60 students took to run 100 metres.

Time, t, seconds	Frequency
$11 < t \leq 13$	8
$13 < t \leq 15$	13
$15 < t \leq 17$	29
$17 < t \leq 19$	10

a Draw a cumulative frequency diagram to show the data.

b Use the diagram to find **i** the median **ii** the interquartile range.

c How many students ran under 12 seconds?

a Add another column to the table and work out the cumulative frequency.

Time, t, seconds	Frequency	Cumulative frequency
$11 < t \leq 13$	8	8
$13 < t \leq 15$	13	21
$15 < t \leq 17$	29	50
$17 < t \leq 19$	10	60

The graph is then drawn by plotting the cumulative frequency against the top point of each group, then joining them with a curve.

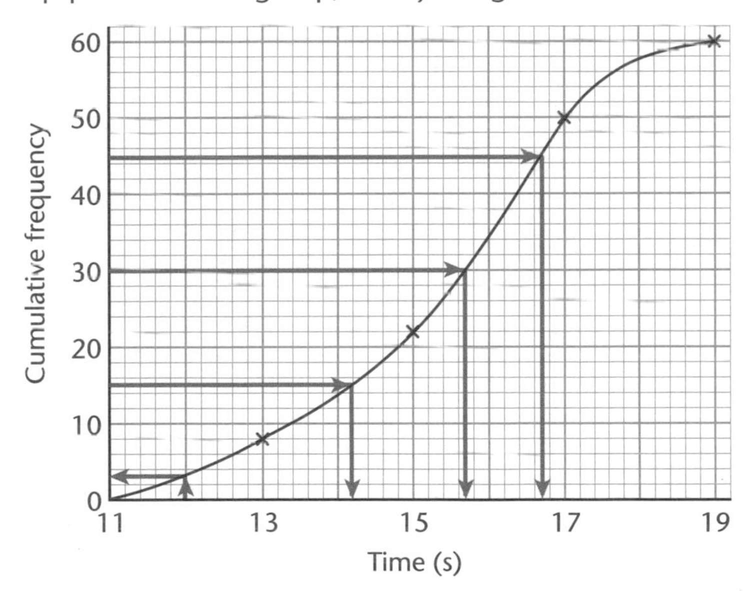

Top Tip!

Always show how you found the median and quartiles by drawing lines on the graph.

b ***i*** The median is found by drawing a line from halfway up the Cumulative frequency axis and reading from the Time axis. So, the median = 15.7 s

ii To find the interquartile range, draw lines across from the quarter and three-quarter values on the Cumulative frequency axis to the graph and then down to the Time axis. This gives the lower and upper quartiles which are subtracted to give the interquartile range.

Upper quartile – lower quartile = 16.7 – 14.2 = 2.5 s.

c Read from 12 on the Time axis to the graph and across to the Cumulative frequency axis. This shows that approximately 3 students ran under 12 seconds.

HANDLING DATA Probability 1

level 6

Probability of events

- The total probability of all possible events is 1.
- These are examples of **mutually exclusive** events:

P(boy) + P(girl) = 1

P(head) + P(tail) = 1.

The probability of an event is $P(\text{event}) = \dfrac{\text{number of ways event can happen}}{\text{number of total outcomes}}$

Example: If the chance of picking a black ball from the bag shown here is $\frac{6}{10}$ or $\frac{3}{5}$, what is the chance of picking a white ball?

You can see that there are four white balls so the chance of picking a white ball is $\frac{4}{10}$ or $\frac{2}{5}$.

Note that $\frac{6}{10} + \frac{4}{10} = 1$ and $\frac{3}{5} + \frac{2}{5} = 1$.

Top Tip!

Unless the question asks for an answer in 'its simplest form', you do not have to cancel fractions – but be careful if you do.

Example: A box contains 21 copper nails and 9 steel nails.

A nail is taken out at random.

a What is the probability that it is a copper nail?
Give your answer as a fraction in its simplest form.

b Give your answer to part **a** as a percentage.

c The first nail is put back in the box. Then six nails are taken out of the box.
After the nails are removed, the probability of taking a copper nail at random is $\frac{7}{8}$.
Explain how you know that the six nails taken out were steel nails.

a $P(\text{copper}) = \frac{21}{30} = \frac{7}{10}$

b $\frac{7}{10} = 70\%$

c There are now 24 nails in the box.
If $P(\text{copper}) = \frac{7}{8} = \frac{21}{24}$, then there are still 21 copper nails in the box.
So the nails removed must have been steel nails.

You need to know the equivalent decimals and percentages for some simple fractions.

oot Check

1 A bag contains 4 red counters and 8 blue counters. A counter is taken out at random.
What is the probability that the counter is
a red **b** blue **c** green?

level 6

Combined events

- Sometimes two separate events can take place at the same time, for example, throwing a dice and tossing a coin.
- The mathematical name for events like this is **independent** because the outcome of throwing the dice does not have any influence on the outcome of tossing the coin.
- The combined outcomes of the two events can be shown in different ways.
- They can be written as a **list**:

  (1, head), (1, tail), (2, head), (2, tail), (3, head), (3, tail),

  (4, head), (4, tail), (5, head), (5, tail), (6, head), (6, tail)
- They can also be shown in a **sample space** diagram.
- You can see that there are 12 outcomes for the **combined events**.

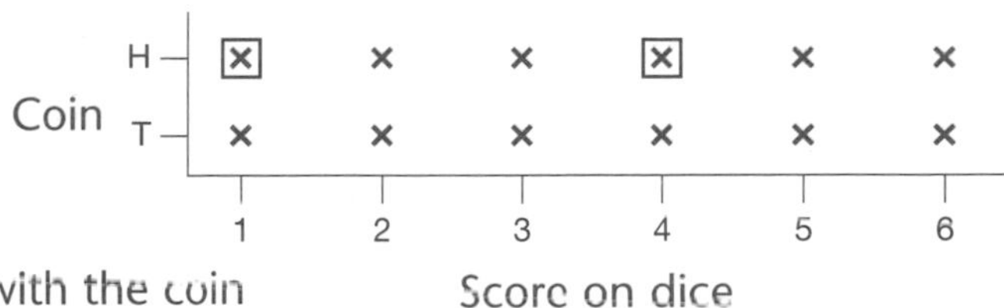

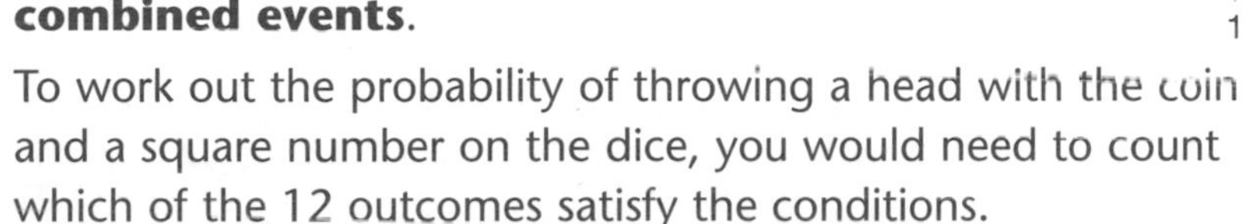

To work out the probability of throwing a head with the coin and a square number on the dice, you would need to count which of the 12 outcomes satisfy the conditions.

These are shown in a box on the sample space diagram.

So P(head and square number) = $\frac{2}{12} = \frac{1}{6}$.

Sample mental test question

There are 6 red and 3 blue balls in a bag. One ball is taken from the bag at random.

What is the probability that it will be blue?

There are three blue balls out of nine altogether. The probability of a blue ball is $\frac{3}{9} = \frac{1}{3}$.

level 6

Sample worked test question

Two four-sided dice numbered from 1 to 4 are thrown together.
The scores are multiplied together.

a Complete the sample space diagram showing the possible scores of the combined event.

b Find the probability that the combined score is

i an even number **ii** a square number **iii** a factor of 144.

Score on second dice (columns); Score on first dice (rows)

	1	2	3	4
1	1	2		
2			6	
3		6	9	
4	4	8	12	16

Answers

a

Score on second dice (columns); Score on first dice (rows)

	1	2	3	4
1	1	2	3	4
2	2	4	6	8
3	3	6	9	12
4	4	8	12	16

b ***i*** *There are 16 outcomes and 12 of them are even numbers. P(even) =* $\frac{12}{16} = \frac{3}{4}$

ii *There are 6 square numbers: 1, 4, 4, 4, 9, 16.*
P(square) = $\frac{6}{16} = \frac{3}{8}$

iii *All of the numbers are factors of 144.*
P(factor of 144) = 1

Did You Know?

If you are in a room with 30 strangers, the chance that one of them has the same birthday as you is over a half.

Probability 2

Tree diagrams

level 8

- Another way of showing **combined events** is to use a **tree diagram**.
- These show **all possible outcomes** of the combined events.

Example: A dice is thrown twice. What is the probability that you get **a** two sixes **b** exactly one six?

This tree diagram shows all the outcomes.

First throw		Second throw		Outcome	
$\frac{1}{6}$	6	$\frac{1}{6}$	6	(6, 6)	$\frac{1}{6} \times \frac{1}{6} = \frac{1}{36}$
		$\frac{5}{6}$	not 6	(6, not 6)	$\frac{1}{6} \times \frac{5}{6} = \frac{5}{36}$
$\frac{5}{6}$	not 6	$\frac{1}{6}$	6	(not 6, 6)	$\frac{5}{6} \times \frac{1}{6} = \frac{5}{36}$
		$\frac{5}{6}$	not 6	(not 6, not 6)	$\frac{5}{6} \times \frac{5}{6} = \frac{25}{36}$

a The probability of two sixes is $\frac{1}{6} \times \frac{1}{6} = \frac{1}{36}$.

b The probability of exactly one 6 is $\frac{1}{6} \times \frac{5}{6} + \frac{5}{6} \times \frac{1}{6} = \frac{10}{36}$.

The probabilities across any branch should add to 1.

Probability of events without replacement

level 8

- Tree diagrams provide a good way of keeping track of probabilities when the items chosen are **not put back**.

Example: A ball is taken from the bag shown and not replaced. Another ball is then taken out.

a The ball chosen is black, what is the probability that the second ball is **i** black **ii** white?

b The ball chosen is white, what is the probability that the second ball is **i** black **ii** white?

c What is the probability that both balls are **i** black **ii** the same colour?

a If the first ball chosen is black, of the 9 balls left, 5 are black and 4 are white.

So, ***i*** P(second ball is black) = $\frac{5}{9}$

ii P(second ball is white) = $\frac{4}{9}$.

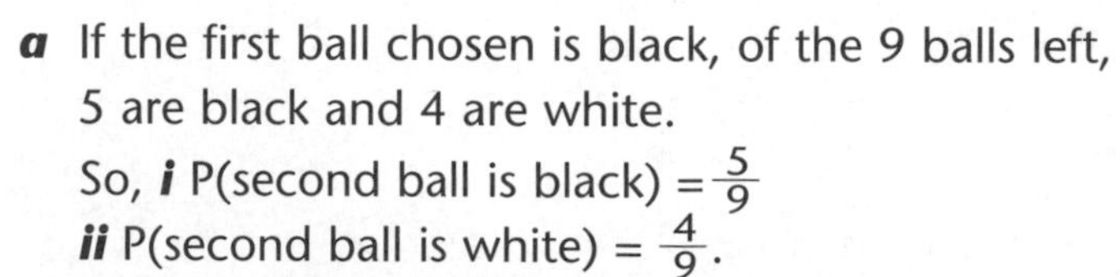

b If the first ball chosen is white, of the 9 balls left, 6 are black and 3 are white.

So, ***i*** P(second ball is black) = $\frac{6}{9} = \frac{2}{3}$

ii P(second ball is white) = $\frac{3}{9} = \frac{1}{3}$.

Probability of events without replacement continued

c This can be done with a tree diagram.

First ball | Second ball | Outcome

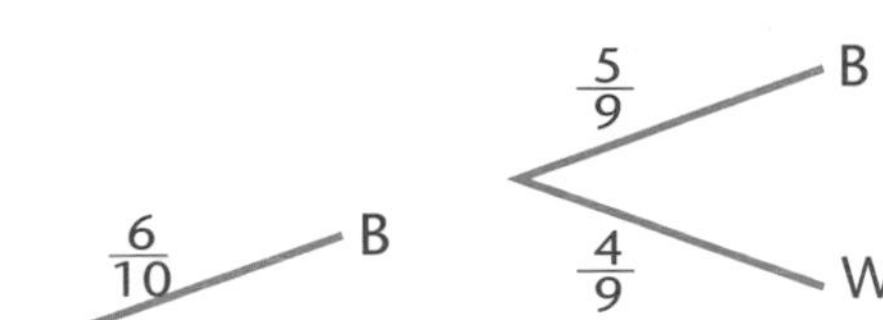

(B, B) $\frac{6}{10} \times \frac{5}{9} = \frac{30}{90}$

(B, W) $\frac{6}{10} \times \frac{4}{9} = \frac{24}{90}$

(W, B) $\frac{4}{10} \times \frac{6}{9} = \frac{24}{90}$

(W, W) $\frac{4}{10} \times \frac{3}{9} = \frac{12}{90}$

Top Tip!

Don't cancel fractions down to start with. Keeping the denominators the same makes it easier if you have to add fractions. Only cancel down the fractions in your final answers.

i P(black and black) = $\frac{6}{10} \times \frac{5}{9} = \frac{30}{90} = \frac{1}{3}$

ii P(same colour) = P(black and black) or P(white and white) = $\frac{30}{90} + \frac{12}{90} = \frac{42}{90} = \frac{7}{15}$

Sample worked test question

level 8

Two children are going to be picked at random from 3 boys and 2 girls.

a Complete the tree diagram to show all the possible outcomes and their probabilities.

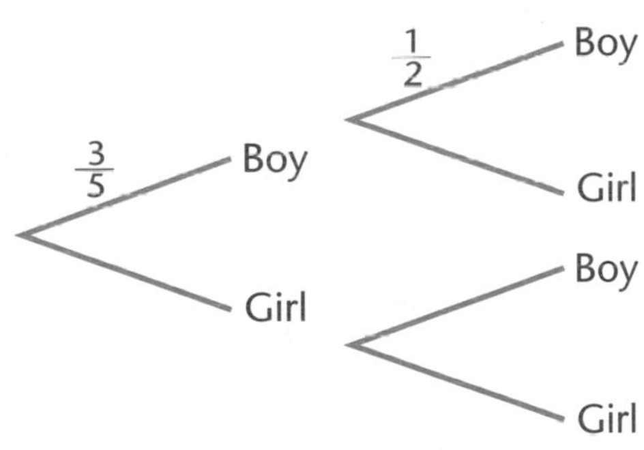

(B, B) $\frac{3}{5} \times \frac{1}{2} = \frac{3}{10}$

b Find the probability that the chosen pair are **i** both girls **ii** a boy and a girl.

Answers

a

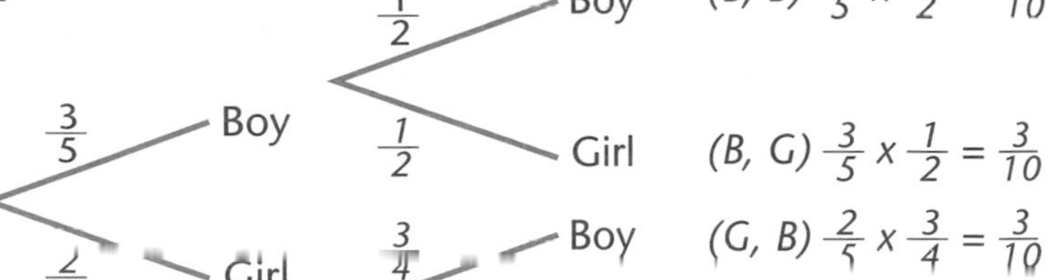

(B, B) $\frac{3}{5} \times \frac{1}{2} = \frac{3}{10}$

(B, G) $\frac{3}{5} \times \frac{1}{2} = \frac{3}{10}$

(G, B) $\frac{2}{5} \times \frac{3}{4} = \frac{3}{10}$

(G, G) $\frac{2}{5} \times \frac{1}{4} = \frac{1}{10}$

b i *P(two girls)* = $\frac{1}{10}$

ii *P(boy and girl)* = $\frac{3}{10} + \frac{3}{10}$

$= \frac{6}{10} = \frac{3}{5}$

Did You Know?

The chances of winning the lottery twice is about 1 in 200 million.

Spot Check

1 A bag contains 4 red counters and 8 blue counters. A counter is taken out at random and not replaced. Another counter is then taken out at random. What is the probability that both counters are **a** red **b** blue?

Index

Answers to Spot Check questions

p. 5 **1 a** 70 **b** 0.00452 **c** 18.72 **d** 1.47

p. 7 **1** 1924 **2** 54

p. 8 **1 a** –9 **b** +9 **c** +12 **d** –5

p. 10 **1 a** $\frac{17}{20}$ **b** $\frac{7}{12}$
2 a $3\frac{11}{12}$ **b** $\frac{29}{40}$

p. 13 **1 a** $\frac{1}{6}$ **b** 3
2 a $\frac{5}{12}$ **b** $1\frac{1}{3}$

p. 15 **1 a** £90 **b** 105.6 kg

p. 16 **1 a** 15% **b** £60
2 32 212

p. 18 **1** 7 : 9
2 £12 : £20

p. 20 **1 a** 216 **b** 2
2 a 7^7 **b** 7^3

p. 22 **1 a** 6000 **b** 6
2 a 365 **b** 374

p. 25 **1 a** 7×10^{-5} **b** 7×10^6
2 a 0.0000062 **b** 5 200 000 000
3 1.28×10^{11}

p. 27 **1** 23, 27 **2** $4n$ **3** $4n + 1$

p. 29 **1 a** odd **b** even **c** odd **d** either

p. 31 **1 a** $10 - 2a$ **b** $20b^2$
2 a 8 **b** 25

p. 33 **1 a** $6a^5b^5$ **b** $6a^3b^2$
2 a $x^2 - x - 6$ **b** $x^2 - 6x + 8$

p. 35 **1 a** $3x(x + 2)$ **b** $2a^2b(2b^2 + 3a)$
2 a $(x - 5)(x - 4)$ **b** $(x + 5)(x - 2)$

p. 36 **1** 6 **2** 5

p. 39 **1**

x	–2	–1	0	1	2	3	4
y	–8	–5	–2	1	4	7	10

p. 41 **1 a** equation iii **b** equation ii **c** equation iv
d equation i

p. 43 **1 a** 6 **b** 10
2 a 4 **b** 7

p. 45 **1 a** 4.5 **b** 7
2 a $x > 2.5$ **b** $x \geq 11$

p. 47 **1** $2 \times (3)^3 + 3 \times 3 = 63$ and $2 \times (4)^3 + 3 \times 4 = 140$

p. 49 **1** $x = 2$, $y = -1$

p. 51 **1 a** 270° **b** 000° or 360° **c** 090° **d** 310° **e** 020°

p. 52 **1** 108°

p. 55 **1** $a = 120°$, $b = 60°$
2 $a = 70°$, $b = 79°$

p. 56 **1 a** A is a reflection in the line $y = 2$
b B is a rotation of 90° clockwise about (2, 1)

p. 59 **1**

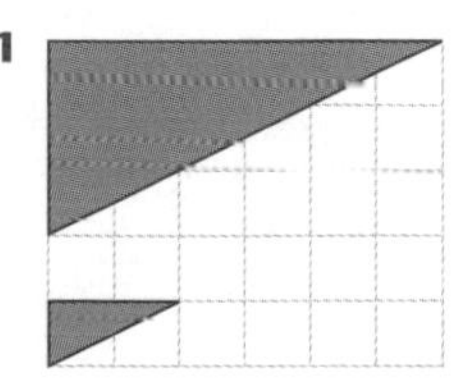

2 If the sides are doubled, the area will be 4 times bigger.

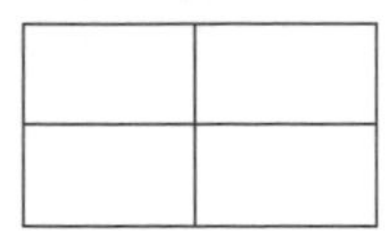

p. 61 **1** A and C

p. 63 **1** Perimeter 24 cm; area 28 cm^2

p. 65 **1** Circumference 18.8 cm; area 28.3 cm^2
2 34.6 m^2

p. 66 **1** Volume 24 cm^3; surface area 52 cm^2

p. 69 **1** Make sure that the sides and angles in your triangle are accurate.

p. 71 **1** Because it is a fixed distance from a point, the shape is a circle.

p. 72 **1 a** Because they have similar angles and the sides are in the same ratio. **b** 8 cm **c** 4.5 cm

p. 75 **1 a** 10.8 cm **b** 12.7 cm

p. 77 **1** 9.65 cm

p. 79 **1 a** 10.5 cm **b** $x = 50°$ because angle at centre is twice angle at circumference.

p. 80 **1 a** mean 7, range 6 **b** mean 7, range 11

p. 83 **1 a** about 5 °C **b** It's not possible to estimate as the temperature may not continue to decrease at the same rate.
2 Between 3 and 4 as the graph is the steepest.

p. 85 **1** 84°
2 a If there were 30 students, each student would be represented by 12° and 20° is not a multiple of 12.
b There could be 18 students, or any multiple of 18. 36 is the most likely number.

p. 87 **1 a** (weak) positive correlation **b** (strong) negative correlation

p. 89 **1** Leading (or offensive) question. Not enough response sections.

p. 90

11 13 15 17 19
Time (s)

p. 92 **1 a** $\frac{4}{12} = \frac{1}{3}$ **b** $\frac{8}{12} = \frac{2}{3}$ **c** 0

p. 95 **1 a** $\frac{1}{11}$ **b** $\frac{14}{33}$

Collins

Collins Revision

KS3

Maths

Workbook

Levels 5–8

Keith Gordon

NUMBER

Multiplying and dividing decimals

level 5

1 The diagram shows how to change metres into millimetres.

Number of metres → $\times 10^2$ → Number of centimetres → $\times 10$ → Number of millimetres

a Change 3.4 metres into millimetres.

____

b Change 74 millimetres into metres.

____

1 mark

2 Work out the following.

a $5.43 \times 10 =$ ________ **b** $0.347 \times 10^2 =$ ________

c $6.72 \div 10 =$ ________ **d** $80.7 \div 10^3 =$ ________

3 Fill in the missing numbers.

a $5 \div 10 =$ ________ **b** ________ $\times 10^2 = 230$

c $0.6 \div 10^2 =$ ________ **d** $0.6 \times$ ________ $= 60$

4 Will spends £17.10 each week in bus fares.

a How much would he spend on bus fares in 4 weeks?

____

b How much would he save with a monthly pass that costs £56.50?

____

5 Rhona bought 3 CDs and 5 DVDs.
How much does she pay altogether?

____

2 marks

level 5

6 Jamal buys 5 new tyres for his car.
The total bill is £325.
How much was each tyre?

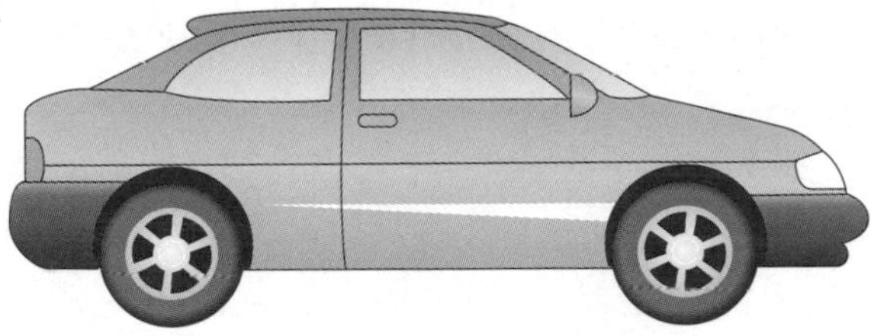

____

1 mark

7 Work out the following.

a 4.6 x 7 = ______

b 35.4 ÷ 6 = ______

c 8 x 5.2 = ______

d 58.1 ÷ 7 = ______

4 marks

8 Six people buy a meal in a restaurant and the bill comes to £256.80.

If they share the cost equally how much does each person pay?

____

1 mark

9 Which is the larger amount:
one-fifth of £46 or one-eighth of £73?

____

1 mark

Long multiplication and division

level 5

1 Work out the following.

a 27 x 32 ________ 2 marks

b 36 x 217 ________ 2 marks

c 952 ÷ 28 ________ 2 marks

d 994 ÷ 14 ________ 2 marks

2 a Eggs are delivered in trays of 48.

How many eggs will be in 17 trays?

________ 2 marks

b A restaurant orders 1000 eggs.

i How many full trays will they need?

________ 2 marks

ii How many eggs will be in the last tray?

________ 2 marks

3 One bus carries 52 passengers.

a How many passengers could be carried on 23 buses?

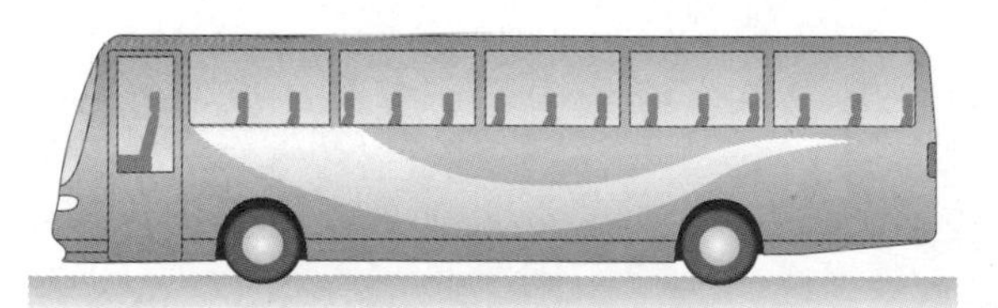

________ 2 marks

b A school is taking 950 students to a theme park at the end of term.

How many buses will they need to hire?

________ 2 marks

4 Maths textbooks are sold in packs of 15.

a A school orders 24 packs.
How many books are there in 24 packs?

________________ **2 marks**

b Another school has 272 students in Year 9.

i How many packs will the school need to order to get a maths textbook for each student?

________________ **2 marks**

ii Will the school have enough spare textbooks to give one to each of the 11 maths teachers?

Show your working. ________________ **1 mark**

5 a Tickets to an orchestral concert in a school hall are £42 each.
350 people attended the concert.
The orchestra charged £12 500 for playing the concert.
How much money did the school make?

________________ **2 marks**

b There were 28 members of the orchestra and one conductor.
Each member of the orchestra was paid £400.
The rest of the money was paid to the conductor.
How much money did the conductor get?

________________ **2 marks**

c The 350 people sat in rows of 24 seats.
How many rows of seats were needed?

________________ **1 mark**

6 What is the remainder when 617 is divided by 23?

________________ **1 mark**

7 How many boxes of cakes, each holding 12 cakes, will be needed to give 120 guests at a garden party three cakes each?

________________ **1 mark**

NUMBER Negative numbers

levels 5-6

1 Look at the following list of numbers.
−7, −6, −2, −1, 0, 2, 4, 8

a What is the total of all eight numbers in the list?

____

b Choose three different numbers from the list that have the lowest total.

____

c Choose two numbers from the list so that the product is as low as possible.

____

2 The diagram shows how to change °C into °F.

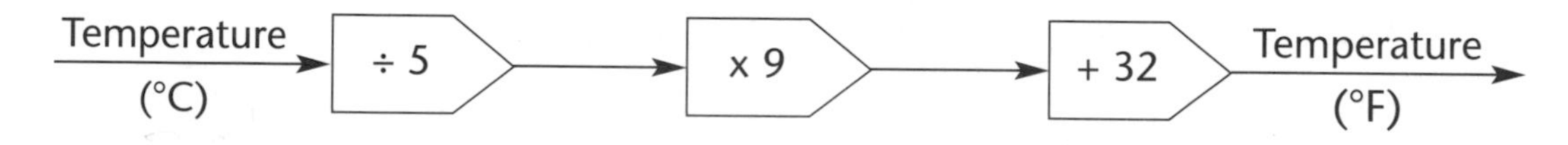

a Change 20 °C into °F. ________________ 1 mark

b Change −40 °C into °F. ________________ 1 mark

c Change −4 °F into °C. ________________ 1 mark

3 Using = (equals), < (less than) or > (greater than), put the correct sign between each number sentence. The first one has been done for you.

a $5 - 6 < 6 - 5$

b −9 ________ −3

c +7 − −8 ________ +8 − −7

d 4 x −2 ________ −4 x 2

3 marks

4 Work out the following.

a −8 + 3 − 6 ________

b −3 x −2 + 5 ________

c −32 ÷ +8 ________

d (−4 − 3) x −6 ________

4 marks

levels 5-6

5 Fill in two **negative** numbers to make the following true.

a ☐ + ☐ = –5

b ☐ – ☐ = –5

2 marks

6 Write the missing numbers on the number lines.

a

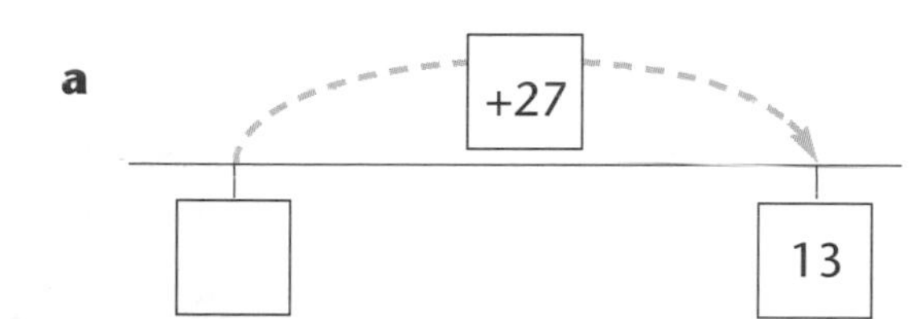

b

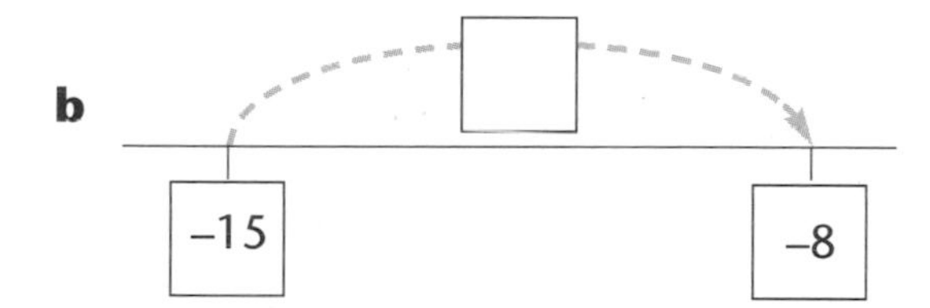

c

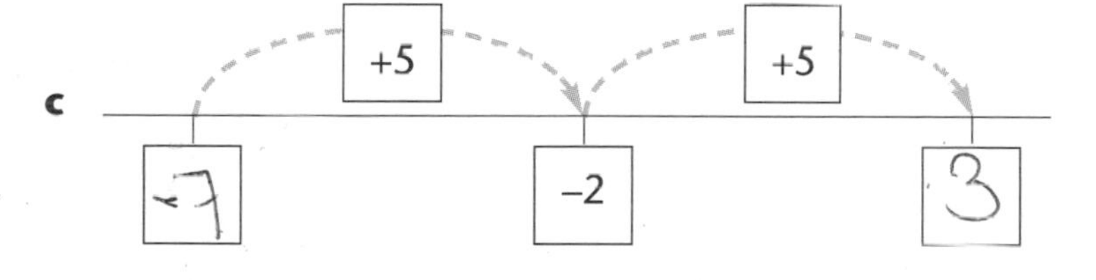

4 marks

7 a Two numbers multiply together to make +12 and add together to make –8. What are the two numbers?

____

2 marks

b 7 squared is 49. The square of another number is also 49. What is the other number?

____

1 mark

8 Look at the following number machine.

a If the input is –3 what is the output? ____________________

1 mark

b If the output is –3 what is the input? ____________________

1 mark

NUMBER

Adding and subtracting fractions

levels 5-6

1 **a** How many sixths are there in $2\frac{5}{6}$? ________________ **1 mark**

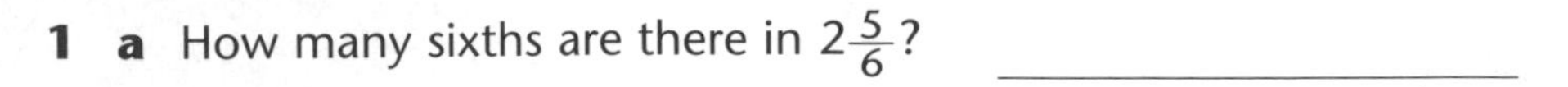

b How many sixths are there in $1\frac{2}{3}$? ________________ **1 mark**

2 **a** Convert $\frac{15}{4}$ to a mixed number. ________________ **1 mark**

b Convert $2\frac{6}{7}$ to a top-heavy fraction. ________________ **1 mark**

3 Work out the following.

a $\frac{1}{3} + \frac{2}{5}$ ________________ **2 marks**

b $\frac{3}{5} - \frac{1}{4}$ ________________ **2 marks**

c $\frac{3}{8} + \frac{3}{5}$ ________________ **2 marks**

d $\frac{7}{8} - \frac{2}{3}$ ________________ **2 marks**

4 Work out the following.

a $2\frac{1}{4} + 3\frac{1}{5}$ ________________ **2 marks**

b $3\frac{2}{5} + 1\frac{1}{8}$ ________________ **2 marks**

c $4\frac{1}{5} - 3\frac{3}{4}$ ________________ **2 marks**

d $2\frac{3}{4} - 1\frac{1}{3}$ ________________ **2 marks**

levels 5-6

5 The diagram shows a grey rectangle that is 8 cm by 6 cm. Two black squares, one 4 cm by 4 cm and the other 2 cm by 2 cm, are drawn inside it.

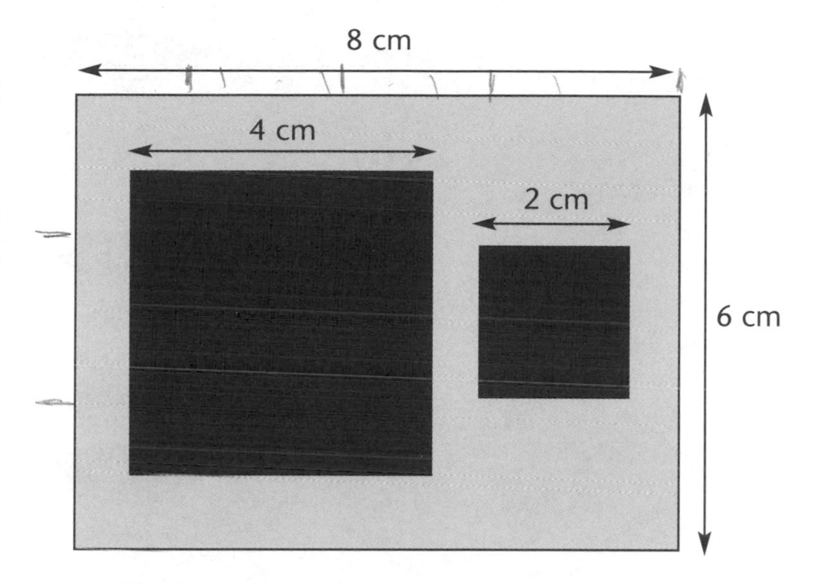

a What fraction of the rectangle is shaded black?

b What fraction of the rectangle is shaded grey?

6 Work out the following.

a $\left(\frac{2}{5}\right)^2$

1 mark

b $\sqrt{\frac{4}{49}}$

c $\sqrt{\frac{4}{9}} \times \left(\frac{3}{5}\right)^2$

NUMBER

Multiplying and dividing fractions

levels 5-6

1 What is $\frac{3}{4} \times \frac{1}{6}$ in its simplest form?

______________ 

2 Work out $\frac{1}{8} \div \frac{5}{6}$. Give the answer in its simplest form.

______________ 1 mark

3 How many $\frac{1}{5}$ are in $2\frac{2}{5}$?

______________ 1 mark

4 **a** Convert $\frac{21}{5}$ to a mixed number. ______________ 1 mark

b Convert $3\frac{1}{6}$ to a top-heavy fraction. ______________ 1 mark

5 Work out the following.

a $\frac{1}{9} \times \frac{3}{5}$ ______________ 1 mark

b $\frac{3}{10} \div \frac{6}{25}$ ______________ 1 mark

c $\frac{3}{5} \times \frac{10}{21}$ ______________ 1 mark

d $\frac{4}{9} \div \frac{2}{3}$ ______________ 1 mark

6 Work out the following.

a $2\frac{1}{4} \times 1\frac{1}{5}$ ______________ 2 marks

b $2\frac{3}{4} \div 4\frac{1}{8}$ ______________ 2 marks

c $3\frac{2}{3} \times 2\frac{1}{4}$ ______________ 2 marks

d $1\frac{7}{8} \div 2\frac{1}{12}$ ______________ 2 marks

7 Work out the area of this rectangle.

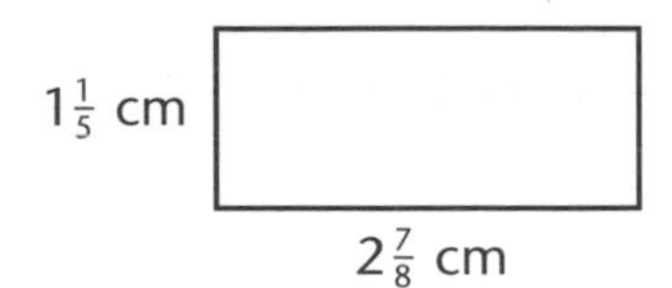

______________ cm^2

8 Work out the areas of these triangles.

a ______________ cm^2

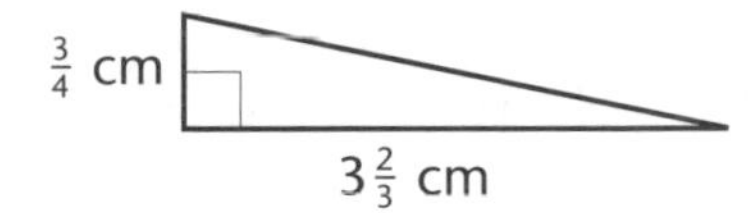

b ______________ cm^2

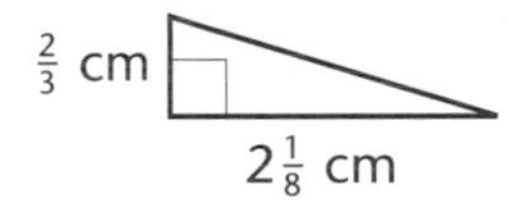

2 marks

9 How many $2\frac{1}{5}$ metre strips of tape can be cut from a roll of tape that is 66 metres long?

____

2 marks

10 This rectangle has an area of 8 cm^2.
Work out the width.

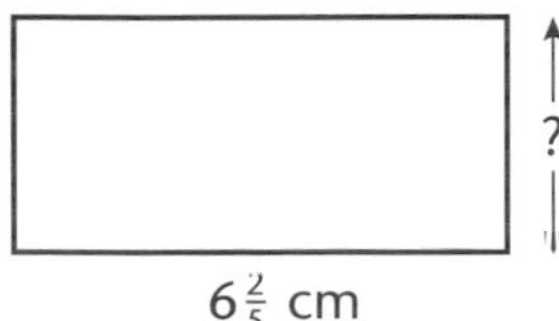

______________ cm

2 marks

NUMBER Percentages 1

levels 5-6

1 Alex asked 50 children what their favourite lunch was.

Lunch	Boys	Girls
Pizza	4	3
Burgers	6	6
Fish cakes	2	3
Sausages	7	6
Salad	1	12
Total	20	30

a What percentage of the children surveyed preferred pizza?

______________ 1 mark

b Which lunch did 10% of the boys prefer? ______________ 1 mark

c Which lunch did 40% of the girls prefer? ______________ 1 mark

d Alex said, 'My survey shows that burgers are just as popular with girls as with boys.' Explain why Alex is wrong.

____

______________________________ 1 mark

e Which lunch is equally popular with boys and girls? ______________ 1 mark

2 A clothes shop is having a sale.
All clothes are reduced by 20%.

a What is the sale price of a jacket normally priced at £60?

______________ 1 mark

b What is the sale price of a shirt normally priced at £32?

______________ 1 mark

c On the last day of the sale, the **sale price** is reduced by a further 10%.
Which of the following is the last day price of a pair of boots normally priced at £100? Tick the correct answer.

☐ £80 ☐ £70 ☐ £72 ☐ £90 1 mark

3 The table shows the 2005 population of each of the world's continents.

Continent	Population (in millions)
Australasia	33
Africa	841
Asia	3825
Europe	735
North America	492
South America	379
World total	6305

a Which continent had about 6% of the world's population in 2005?

____

b In 2005, what percentage of the world's population was living in Africa?

____

2 marks

4 The pie chart shows how a farmer uses his land.
The angle for fallow land is 45°.

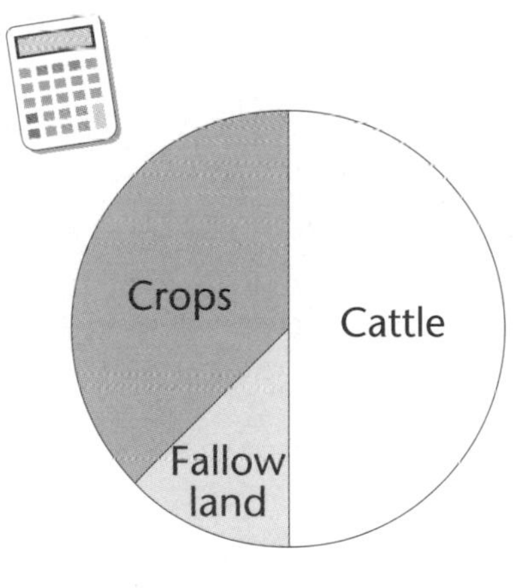

a What percentage of the farm is fallow land?

____

b 330 acres are used for crops. What is the total acreage of the farm?

____

2 marks

c Next year the farmer plans to decrease the acreage for cattle by $33\frac{1}{3}$% and increase the amount for crops by $33\frac{1}{3}$%.
Tick the statement that is true.

- ☐ The amount of fallow land will stay the same.
- ☐ The amount of fallow land will increase.
- ☐ The amount of fallow land will decrease.
- ☐ You cannot tell how the amount of fallow land will change.

1 mark

Percentages 2

level 7

1 In 2004, 320 passengers on average used the train line between Barnsley and Huddersfield each weekday.

In 2005, 464 passengers on average used the same train.

What percentage increase is this?

_______________ % **2 marks**

2 A standard box of cereal contains 900 g of cereal.

It is recommended that one serving is 60 grams.

a How many servings will you get from a standard box?

_______________ **1 mark**

b A special offer box contains 20% more cereal than a standard box.

How many servings will you get from this box?

_______________ **1 mark**

3 **a** What percentage of 25 is 4?

_______________ % **1 mark**

b What percentage is 6 of 48?

_______________ % **1 mark**

4 Robert got the following marks for a series of tests.

Put them in percentage order, with the smallest percentage first.

6 out of 10; 13 out of 20; 40 out of 64; 32 out of 50.

____ **2 marks**

5 After pruning, a conifer was decreased in height from 4 m 40 cm to 3 m 74 cm.

What percentage decrease is this?

_______________ % **2 marks**

6 A Petri dish contains 35 000 bacteria.

A disinfectant is added that kills 12% of the bacteria each minute.

How many bacteria will there be after 10 minutes?

_______________ **2 marks**

level 7

7 An ant colony increases by 6% per day. 

Initially there are 1000 ants.

a How many will there be after a week?

__________________ 2 marks

b After a week the colony is treated with an insecticide that decreases the population by 6% per day.

How many ants will there be after a further week?

8 After a 12% reduction the price of a TV is £308.

What was the original price of the TV?

2 marks

9 After treating with fertilizer the average weight of John's tomatoes increased by 15% to 92 grams. 

What was the average weight before the treatment?

__________________ g

10 This rectangle has an area of 24 cm^2.

The length increases by 10%.

The width increases by 20%.

What is the percentage increase of the area?

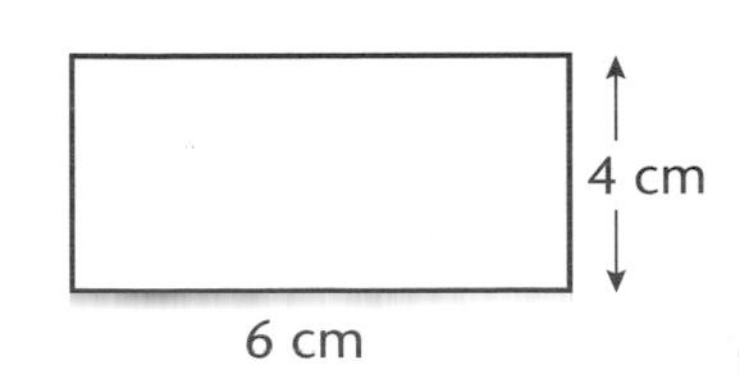

__________________ % 2 marks

NUMBER Ratio

levels 5-6

1 **a** Write the ratio 6 : 9 in its simplest form. __________ 1 mark

b Write the ratio 15 : 25 in its simplest form. __________ 1 mark

2 The ratio of two packets of cornflakes is 3 : 4

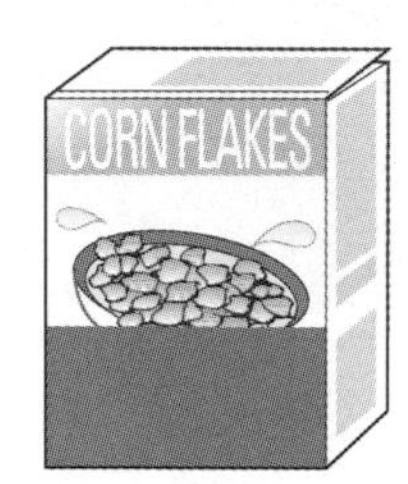

Find the mass of the larger packet. __________

3 Shade the diagram so that the ratio of shaded squares to unshaded squares is 1 : 3

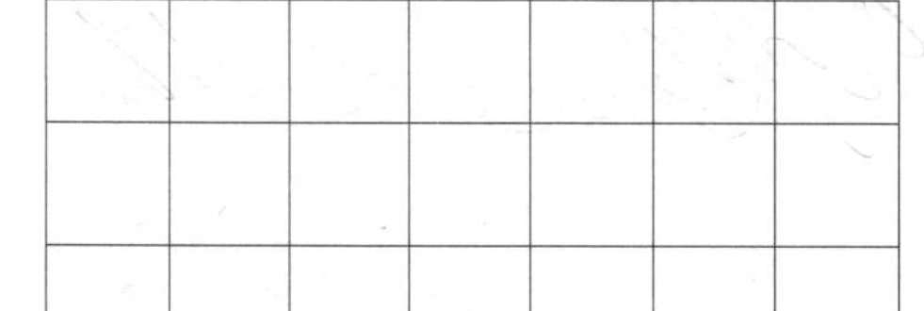

4 **a** Divide £90 in the ratio 1 : 4

____

b Divide 150 kg in the ratio 2 : 3

__________ 2 marks

5 A drink is made from cranberry juice and lemonade in the ratio 2 : 7

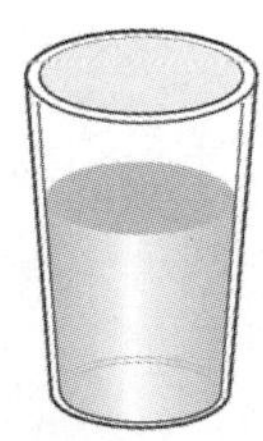

a How much lemonade is needed if 50 ml of cranberry juice is used?

__________ 2 marks

b How much cranberry juice is in 450 ml of the drink?

__________ 2 marks

6 The diagram shows a grey rectangle 8 cm by 6 cm with a black square 3 cm by 3 cm drawn inside it.

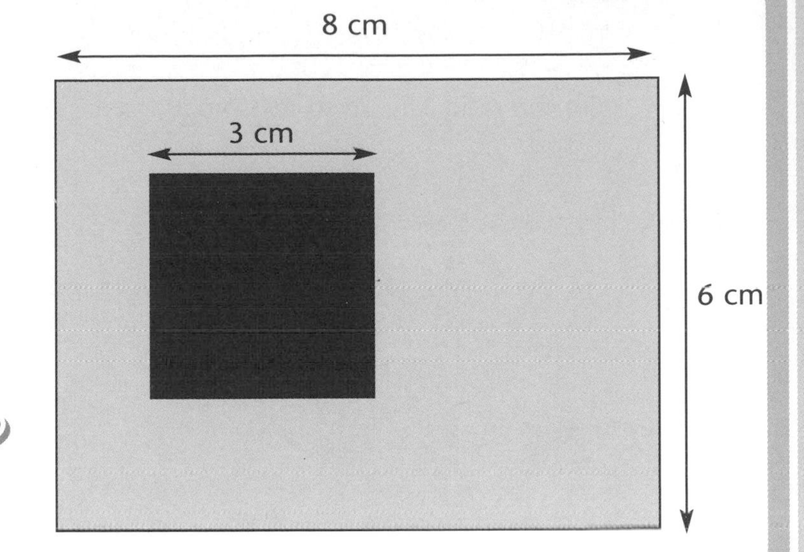

a Calculate the ratio of the perimeter of the rectangle to the perimeter of the square. Give your answer in its simplest form.

______________________ **2 marks**

b Calculate the ratio
grey area : black area
Give your answer in its simplest form.

______________________ **2 marks**

7 Fatima won £146 on the lottery.

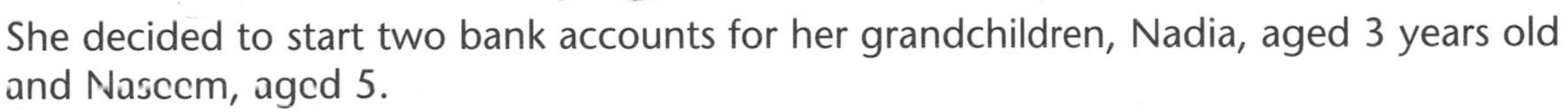

She decided to start two bank accounts for her grandchildren, Nadia, aged 3 years old and Naseem, aged 5.
She shared the money between the children in the ratio of their ages.

a How much did each child get in the bank? ______________________ **2 marks**

b The following year she won another £146 and did the same thing with the money.
How much would each child have in the bank in total now?

______________________ **2 marks**

8 A fizzy drink is sold in two sizes. The small bottle costs 35p and the larger bottle costs 60p.

a Write down the ratio of the sizes of the bottles in its simplest form. ______________________

b Write down the ratio of the costs of the bottles in its simplest form. ______________________ **1 mark**

c Which size is the best value for money?
Explain your answer. ______________________

______________________ **2 marks**

NUMBER Powers and roots

levels 6-7

1 Work out the following.

a 2^6 ______ 1 mark

b 3^7 ______ 1 mark

2 Work out the following leaving the answer in index form.

a $4^4 \times 4^3$ ______ 1 mark

b $5^8 \div 5^2$ ______ 1 mark

3 Put these numbers in order of size, smallest first.

$\sqrt[3]{64}$ 2^3 3^2 $\sqrt{25}$ 25°

______ 2 marks

4 Here are some number cards.

2^4 4^2 3^3 8^1 8^0

a Which number is the largest? ______ 1 mark

b Which two cards have the same value? ______ 1 mark

5 Some numbers are smaller than their squares.

For example: $3 < 3^2$

a Which number is equal to its square? ______ 1 mark

b Give an example of a number that is greater than its square. ______ 1 mark

6 $\sqrt{30}$ lies between 5 and 6 because $5^2 = 25$ and $6^2 = 36$.

Find two whole numbers between which the following must lie.

a $\sqrt{15}$ ______ 1 mark

b $\sqrt[3]{10}$ ______ 1 mark

levels 6-7

7 Find values of a, b and c such that

$64 = 2^a = 4^b = 8^c$

$a =$ ______

$b =$ ______

$c =$ ______

8 Find the values of m and n such that

$2^m \times 3^n = 108$

$m =$ ______

$n =$ ______

9 A cube has a volume of 125 cm^3.

a What is the length of the side?

______ cm

b What is the surface area of the cube?

______ cm^2

10 Shona writes down that $\sqrt{x + y} = \sqrt{x} + \sqrt{y}$

Give an example to show that Shona is wrong.

______

11 Tony thinks that $\sqrt{x \times y} = \sqrt{x} \times \sqrt{y}$

Give an example to show that Tony is correct.

______

12 There is only one pair of numbers, a and b, which are different and for which $a^b = b^a$

What values are a and b?

______

1 mark

Approximations and limits

levels 7-8

1 Round off the following numbers to 1 significant figure.

a 46.89 ____________ **b** 0.0065 ____________ 2 marks

2 Complete the following table.

Number	Rounded to 1 s.f.	Rounded to 2 s.f.	Rounded to 3 s.f.
5.682			
34 639			
0.09938			

9 marks

3 The width of a square tile is 150 mm, correct to the nearest millimetre.

a What is the least possible width of one tile? ____________ mm 1 mark

b What is the greatest possible width of one tile? ____________ mm 1 mark

c 12 tiles are placed together.
What is the least possible width of the 12 tiles? ____________ mm 1 mark

4 The weight of a chocolate bar is 40 grams to 1 significant figure.

a What is the least possible weight of one chocolate bar? ________ g 1 mark

b What is the greatest possible weight of one chocolate bar? ________ g 1 mark

c Gita has four chocolate bars.
What is the greatest possible weight of Gita's chocolate bars? ________ g 1 mark

5 This table shows the percentage of nationalities on an aeroplane.

Nationality	British	American	French	German	Italian
Percentage	32.5	19.6	22.1	15.1	10.7

a Add up all the percentages. ________ % 1 mark

b Round all the percentages to the nearest whole number. ____________ 1 mark

c Add up the rounded values. Explain your answer. ________ %

____ 1 mark

levels 7-8

6 Work out an approximate answer to the following.

a $\dfrac{412 \times 39}{18.3}$

________________ 1 mark

b $\dfrac{21.8 \times 42.6}{9.25 \times 5.24}$

____

7 Use your calculator to work out the following.

$\dfrac{2.51 \times 6.62}{\sqrt{4.84}}$

a Write down all the digits on your calculator display.

________________ 1 mark

b Round off your answer to 2 significant figures.

________________ 1 mark

8 A rectangle has a length of 8 cm and a width of 6 cm, with both values measured to the nearest centimetre.

a What are the least and greatest values of the length?

least __________ cm greatest __________ cm

b What are the least and greatest values of the width?

least __________ cm greatest __________ cm

c What are the least and greatest values of the area?

least __________ cm^2 greatest __________ cm^2

9 A cube has a side of 5 cm measured to the nearest centimetre.

What is the greatest possible volume?

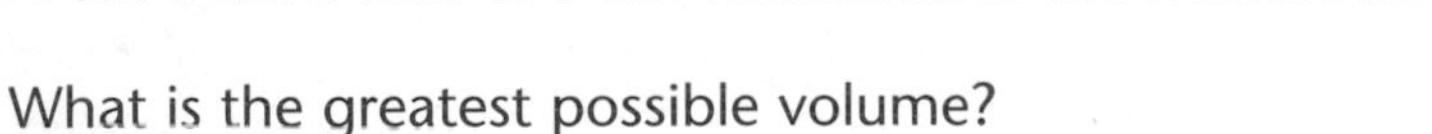

________________ cm^3 1 mark

NUMBER Standard form

level 8

1 Write the following numbers in standard form.

a 0.000008 ____________ **1 mark**

b 67 000 000 000 ____________ **1 mark**

2 Complete the following table, writing the rounded numbers in standard form.

Number	Rounded to 1 s.f.	Rounded to 2 s.f.	Rounded to 3 s.f.
0.004578			
34 640 000			
0.00009638			

9 marks

3 Write the following standard form numbers as ordinary numbers.

a 6.8×10^5 ____________ **1 mark**

b 8.9×10^{-4} ____________ **1 mark**

c 9.85×10^8 ____________ **1 mark**

4 The following numbers are not in standard form.
Write them correctly in standard form.

a 34.5×10^6 ____________ **1 mark**

b 0.7×10^5 ____________ **1 mark**

c $4.2 \times 10^6 \times 5 \times 10^2$ ____________ **1 mark**

d $4.2 \times 10^6 \div 5 \times 10^2$ ____________ **1 mark**

5 Write the following numbers in standard form.

a 8 million ____________ **1 mark**

b one thousandth ____________ **1 mark**

c sixteen thousand ____________ **1 mark**

level 8

6 The Earth is an approximate sphere with a radius of 6400 km.

a The formula for working out the volume of a sphere is

$$V = \frac{4}{3}\pi r^3$$

Work out the volume of the Earth, giving your answer in standard form to 2 s.f.

____

b The surface area of a sphere is given by

$$A = 4\pi r^2$$

70% of the surface area of the Earth is covered by water. Work out the area of water, giving your answer in standard form to 2 s.f.

____

7 A nanometre (nm) is 1×10^{-9} of a metre.
A hydrogen atom is 0.12 nm across.

Write this value in metres in standard form.

____

8 Computer screens are split into pixels. One screen has a resolution of 72 pixels per inch. There are 2.54 centimetres to an inch.

a Using the above information, write down the width of a pixel in centimetres. Give your answer in standard form.

______________________ cm

b Pixels are square in shape. What is the area of an individual pixel? Give your answer in square centimetres in standard form to 3 s.f.

______________________ cm^2

c One screen is 800 pixels by 600 pixels. What is the area of the screen? Give your answer in square centimetres to 1 d.p.

______________________ cm^2

1 mark

ALGEBRA Sequences

level 6

1 The table shows the counting numbers arranged in a six column grid.

	Col 1	Col 2	Col 3	Col 4	Col 5	Col 6
Row 1	1	2	3	4	5	6
Row 2	(7)	8	9	10	11	12
Row 3	13	(14)	15	16	17	18
Row 4	19	20	(21)	22	23	24

a Which column will the number 37 be in?

________________ 1 mark

b Column 2 makes the sequence 2, 8, 14, 20, ...
What will be the 10th term of this sequence?

____

c The nth term of the numbers in the 4th column is $6n - 2$.
What is the nth term of the numbers in the 5th column?

____

d The 7 times table is circled. Row 1 does not have any numbers in the 7 times table.

Which is the next row without a number in the 7 times table?

________________ 1 mark

2 What is the next number in this sequence?

10, 7, 4, 1, –2, ... ________________ 1 mark

3 A sequence has the term-to-term rule 'multiply by 2 and add 1'.
Which of these series of three terms could be in the sequence?
Tick the correct answer. (There may be more than one.)

- ☐ 1, 3, 5, ...
- ☐ 2, 5, 11, ...
- ☐ 10, 21, 63, ...
- ☐ –3, –5, –9, ...

1 mark

4 The nth term of a sequence is $2n - 1$.
Write down the first three terms in the sequence. ________________ 2 marks

level 6

5 Look at this series of patterns.

Pattern 1

Pattern 2

Pattern 3

Pattern 4

a How many grey hexagons will there be in Pattern 6?

1 mark

b What is the nth term of the sequence of white hexagons?

1 mark

6 Look at this series of patterns.

Pattern 1

Pattern 2

Pattern 3

Pattern 4

What is the nth term of the number of hexagons in each sequence?

2 marks

7 What is the missing term in this sequence?

3, 4, ... , 12, 19, 28, 39, 52

1 mark

8 What is the nth term in this sequence?

8, 12, 16, 20, 24, ...

2 marks

Square numbers, primes and proof

levels 5-8

1 Write down the factors of 24. ______________________ 2 marks

2 What is the highest common factor of 24 and 64?

______________________ 1 mark

3 What is the lowest common multiple of 9 and 12?

______________________ 1 mark

4 Which of the following is both a square number and a triangle number?
Tick the correct answer.

☐ 4 ☐ 9 ☐ 25 ☐ 36 1 mark

5 Write down a square number between 101 and 149. ______________________ 1 mark

6 Here are 10 number cards. 1 2 3 4 5 6 7 8 9 10

From the cards, write down

a the square numbers ______________________ 2 marks

b the prime numbers ______________________ 2 marks

c the factors of 10 ______________________ 1 mark

7 Circle A contains the first ten multiples of 2.
Circle B contains the first seven multiples of 3.

Write down the missing numbers from the overlap.

____

____

______________________ 2 marks

levels 5-8

8 Circle A contains the factors of 20.
Circle B contains the factors of 36.

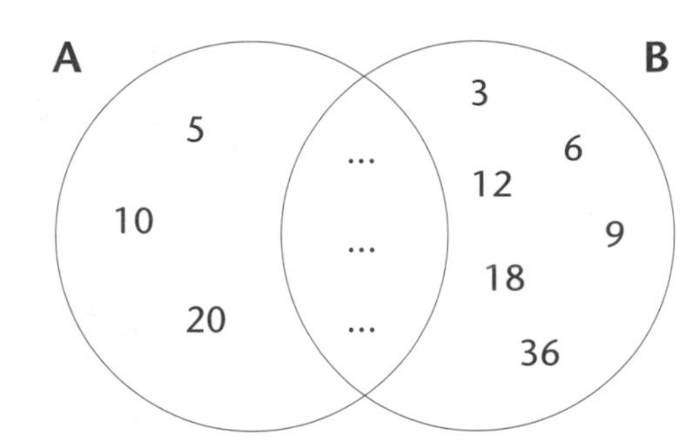

Write down the missing numbers from the overlap.

______

______

______

2 marks

9 p is a prime number, q is an even number, r is an odd number.
State whether the following expressions are:
Always even, Always odd, Could be either odd or even.

		Even	Odd	Either
a	p^2	☐	☐	☐
b	$p(q + r)$	☐	☐	☐
c	q^2	☐	☐	☐
d	pqr	☐	☐	☐

4 marks

10 n is an integer.

a Explain why $2n$ is an even number.

____

1 mark

b Explain why $2n + 1$ is an odd number.

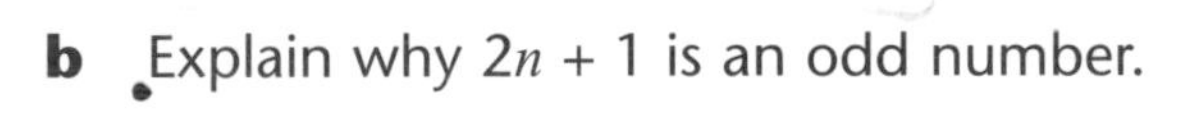

____

1 mark

c Prove that the product of an even number and an odd number is always even.

1 mark

11 Prove algebraically that the sum of three consecutive integers is a multiple of 3.

2 marks

Algebraic manipulation 1

level 5

1 If $a = -3$, $b = +4$ and $c = -5$ which of the following expressions is equal to 3?
Tick the correct answer. (There may be more than one.)

☐ $a(b + c)$ ☐ $a^2 + b^2$

☐ $ab + ac$ ☐ $(c - b) \div a$

1 mark

2 Simplify the following expressions.

a $4(2x - 3)$ ________ 1 mark

b $3a \times 5a$ ________ 1 mark

c $3(a + 2b) + 4(2a - b)$ ________ 1 mark

d $5(3x + 2) - 5(2x - 3)$ ________ 1 mark

3 William has a pile of cards.
The total number of cards is $4n + 8$.

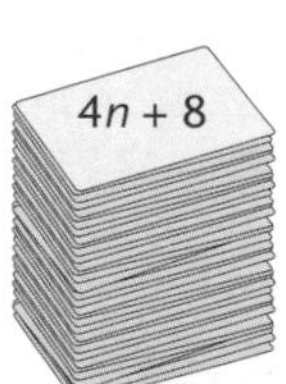

a William puts the cards into 2 piles.
The number of cards in one pile is $3n + 1$.
How many cards are in the other pile?

________ 1 mark

b William puts the cards into 4 equal piles.

How many cards are in each pile?

________ 1 mark

c William counts the cards and finds he has 32 in total.
What is the value of n?

________ 1 mark

4 Look at the rectangle. Write down expressions for the lengths marked **a** and **b**.

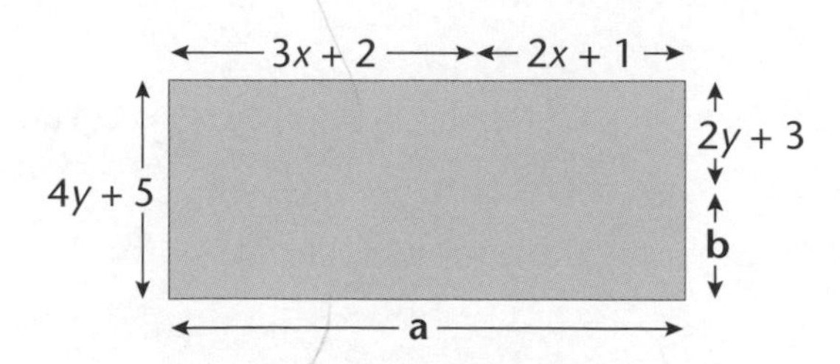

a = ________ 1 mark

b = ________ 1 mark

level 5

5 Here are four cards with algebraic expressions on them.

$2(x + 1)$	$3(x - 2)$	$2(x - 1)$	$3(x + 1)$
Card A	Card B	Card C	Card D

Work out the algebraic expressions formed by:

a Card A + Card B ______ **1 mark**

b Card A + Card C ______ **1 mark**

c Card B − Card C ______ **1 mark**

d Card D − Card A ______ **1 mark**

6 In these walls each brick is made by adding together the two bricks below it.
For example:

$2x + 4$	
$2x$	4

Write the missing expressions in the walls below as simply as possible.

a

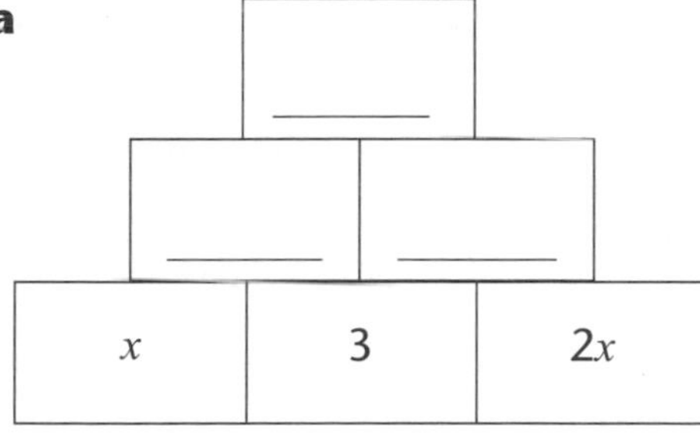

b

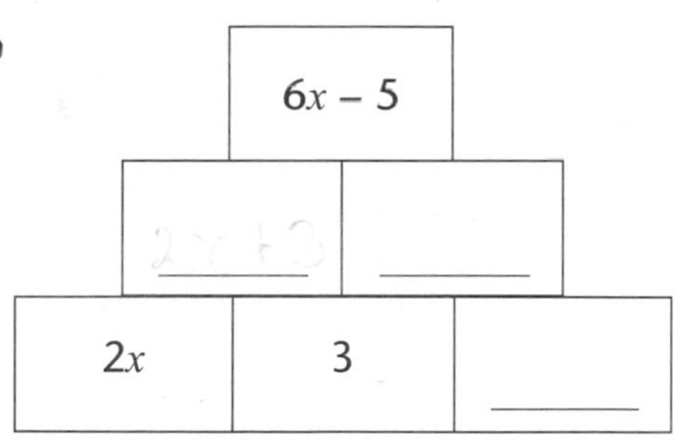

6 marks

7 Work out **a** the perimeter and **b** the area of this shape in terms of x.

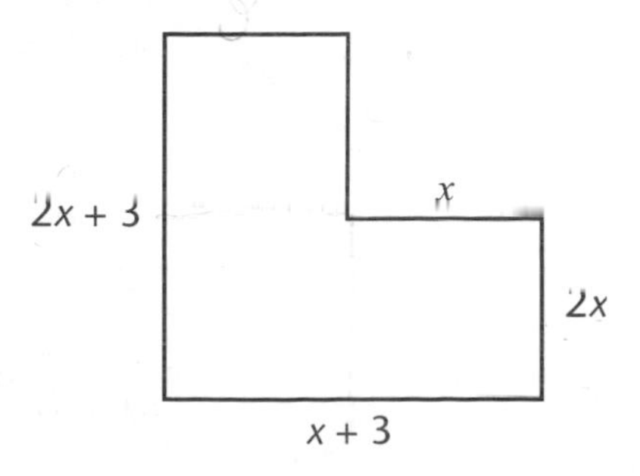

a Perimeter ______ **1 mark**

b Area ______ **1 mark**

Algebraic manipulation 2

levels 6-7

1 Simplify the following expressions.

a $3a^2b \times 4ab$ ______________ 1 mark

b $2a^3b^2 \times 5a^2b^3$ ______________ 1 mark

c $\dfrac{6a^2b^3 \times 4a^2b}{3a^3b^2}$ ______________ 1 mark

d $\dfrac{6ab^3 \times 4a^3b}{2a \times 3b}$ ______________ 1 mark

2 Expand the following brackets.

a $(x - 3)(x + 4)$ ______________ 1 mark

b $(x + 5)(x + 3)$ ______________ 1 mark

c $(x - 1)(x - 2)$ ______________ 1 mark

d $(x + 2)(x - 4)$ ______________ 1 mark

3 a Complete the boxes to expand $(2x + 5)(x - 2)$.

	$2x$	$+ 5$
x	$2x^2$	$+ 5x$
-2	______	______

$(2x + 5)(x - 2) = 2x^2 + 5x$ ______________

= ______________ 1 mark

b Expand the following brackets.

i $(2x + 3)(x + 4)$ ______________ 1 mark

ii $(3x - 1)(2x + 1)$ ______________ 1 mark

iii $(4x + 2)(3x - 4)$ ______________ 1 mark

levels 6-7

4 The rectangle is $(n + 5)$ cm long and $(n + 3)$ cm wide.
It has been split into four smaller rectangles.

a Find the area of each small rectangle.

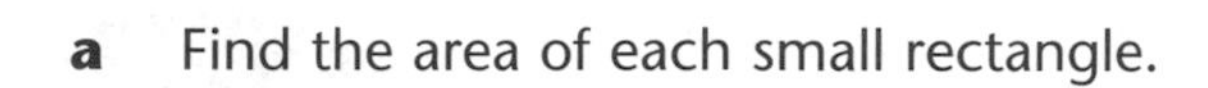

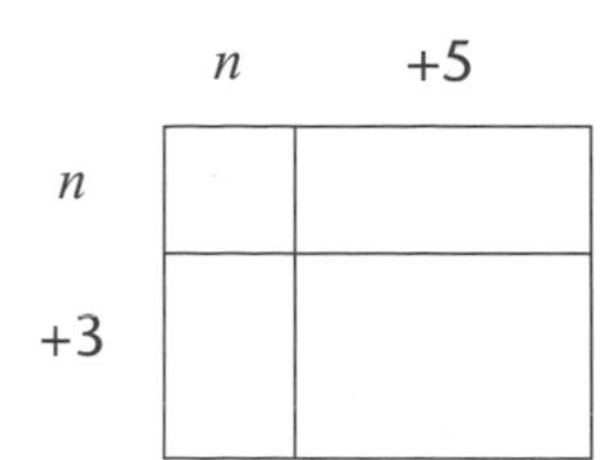

1 mark

b Write down the expansion of $(n + 5)(n + 3)$.

____

2 marks

5 Expand the following brackets.

a $(x - 3)(x + 3)$ ______________________ 1 mark

b $(x + 5)(x - 5)$ ______________________ 1 mark

c $(x - 1)(x + 1)$ ______________________ 1 mark

d $(x + 4)(x - 4)$ ______________________ 1 mark

6 Show that the area of this triangle is $p^2 + 6p + 5$.

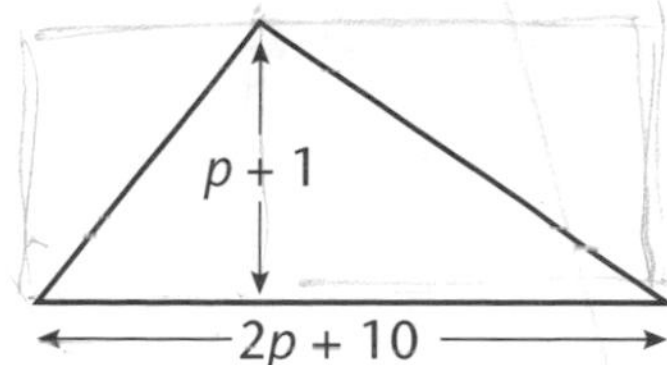

____

2 marks

7 Work out the area of this shape in terms of x.

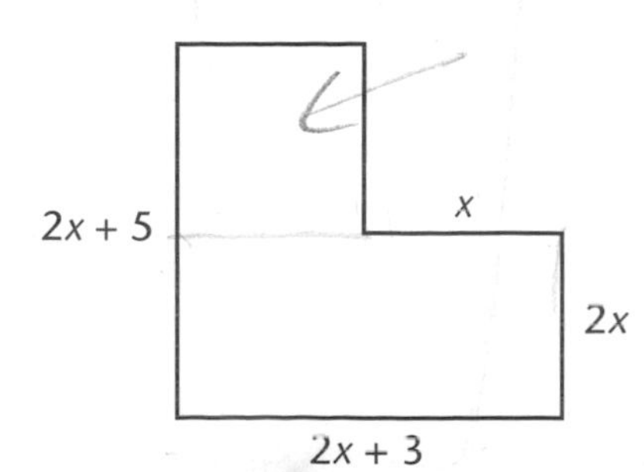

Area = ______________________

2 marks

Factorisation

levels 7-8

1 Which of the following are **not** factors of both $12a^2b^3$ and $18ab^2$?
Tick the correct answers.

☐ $3ab$ ☐ $2a$ ☐ $18ab$ ☐ $4b$ ☐ $6ab^2$

1 mark

2 Which of the following are factors of both $15x^3y^4$ and $25xy^3$?
Tick the correct answers.

☐ $5xy$ ☐ $10x$ ☐ $5x^2y$ ☐ x^3y^4 ☐ $5xy^2$

1 mark

3 Which two of the expressions below are equivalent?

☐ $4(3a - 5)$ ☐ $3(4a - 5)$ ☐ $2(6a - 10)$ ☐ $4(a^3 - 5)$

1 mark

4 a Which of the expressions below is equivalent to the expression

$6x^2y^2 + 24x^3y$

$6xy(xy + 4x^2y)$ $6xy(xy^2 + 4x^2)$

$2xy(3xy + 12x^2)$ $3xy(2xy^2 + 8x^2)$

1 mark

b Which of the expressions below is equivalent to the expression

$x^2 + 8x + 12$

$(x + 3)(x + 4)$ $(x - 3)(x - 4)$

$(x + 2)(x + 6)$ $(x - 2)(x - 6)$

1 mark

5 Which of these expressions is the odd one out?
Explain your answer.

☐ $2y^2(y - 6)$ ☐ $2y^3 - 12y^2$ ☐ $2y(y^2 - 10)$ ☐ $2y(y^2 - 6y)$

____

2 marks

6 a Factorise the following expression.

$5x + 25$ __________________

1 mark

b Factorise the following expression as fully as possible.

$12x^3 - 4x^2$ __________________

1 mark

levels 7-8

7 Factorise the following expressions.

a $3a^2b + 4ab$ ____________ 1 mark

b $12a^3b^2 + 4a^2b^3$ ____________ 1 mark

c $6a^2b^3 + 4a^2b$ ____________ 1 mark

d $6a^2b^3 + 4a^3b^2$ ____________ 1 mark

8 Complete the following factorisations.

a $x^2 - x - 12 = (x -$ ________ $)(x + 3)$ 1 mark

b $x^2 - 8x + 15 = (x - 5)(x$ ________ $)$ 1 mark

c $x^2 + 3x + 2 = (x +$ ________ $)(x +$ ________ $)$ 1 mark

d $x^2 + 2x - 24 = (x +$ ________ $)(x -$ ________ $)$ 1 mark

9 **a** Use the diagram to work out the length and the width of the rectangle.

← length →

← width →

n^2	$2n$
$5n$	10

____________ 1 mark

b Factorise $x^2 + 3x - 4$.

____________ 1 mark

Formulae

levels 5-8

1 The flow diagram shows a formula.

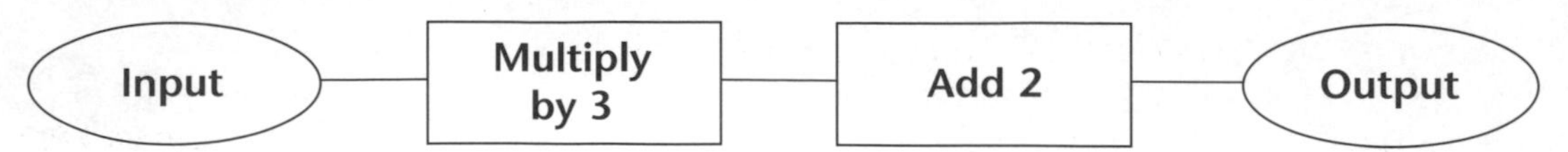

Which of the following pairs of inputs and outputs work for this formula?
Tick the correct answer. (There may be more than one.)

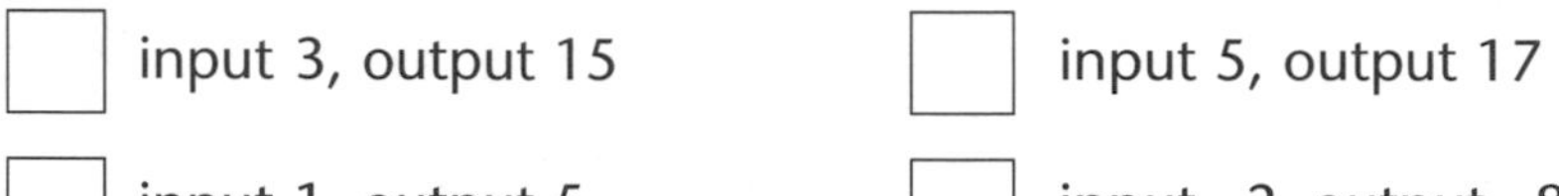

☐ input 3, output 15 ☐ input 5, output 17

☐ input 1, output 5 ☐ input –2, output –8

2 Tim is thinking of a number.

I think of a number.
I double it and add 7.
The answer is 12.

What was the number Tim thought of? ____________

1 mark

3 Input — Multiply by 4 — Add 1 — Output

a What is the output for this flow diagram if the input is 5? ____________

b What is the input for this flow diagram if the output is 5? ____________

4 x — – 2 — x 3 — Output

What formula will be the output from this flow diagram? ____________

5 In a sale the prices are reduced by 20%.
This flow diagram shows how to work out the sale price.

Original price — x 0.8 — Sale price

a What is the sale price of an item with an original price of £20?

____________ 1 mark

b What is the original price of an item with a sale price of £32?

____________ 1 mark

levels 5-8

6 A formula connecting three variables x, y and z is $x = y(8 - z)^2$.  

a Work out the value of y when $x = 10$ and $z = 3$. __________

b Work out the **values** of z when $x = 16$ and $y = 4$. __________

7 The following two equations $v = u + at$ and $v^2 = u^2 + 2as$, connect 

s, distance in metres

u, initial speed in metres per second

v, final speed in metres per second

a, acceleration in metres per second squared

t, time of journey.

For a particular journey, $v = 25$ m/s, $u = 10$ m/s, $t = 2$ seconds.

Use the equations to find s, the distance travelled.

____

8 The shape shown consists of two cylinders fixed together.
The larger cylinder has a radius of a cm and a height of b cm.
The smaller cylinder has a radius of c cm and a height of a cm.
The volume and surface area of the shape are given by the formulae

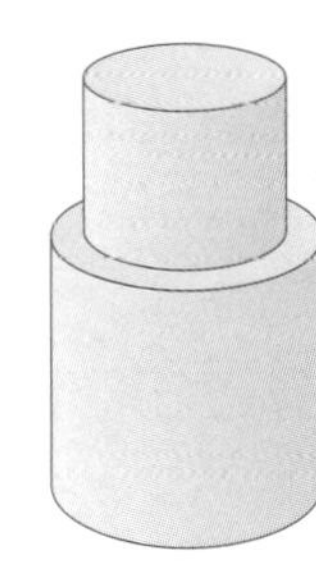

$$V = \pi a(ab + c^2)$$

$$A = 2\pi a(a + b + c)$$

a Work out the volume and area if $a = 10$ cm, $b = 20$ cm and $c = 6$ cm.

Leave your answers in terms of π.

Volume = ____________________

Area = ____________________

2 marks

b Work out the volume and area if $a = 5.6$ cm, $b = 12.2$ cm and $c = 4.8$ cm. 

Give your answers to 3 significant figures.

Volume = ____________________

Area = ____________________

2 marks

Graphs 1

levels 5-6

1 Some rectangular tiles are placed on a coordinate grid.

On the first tile the corner marked with a square has the coordinate (2, 2) and the corner marked with a triangle has the coordinate (0, 1).

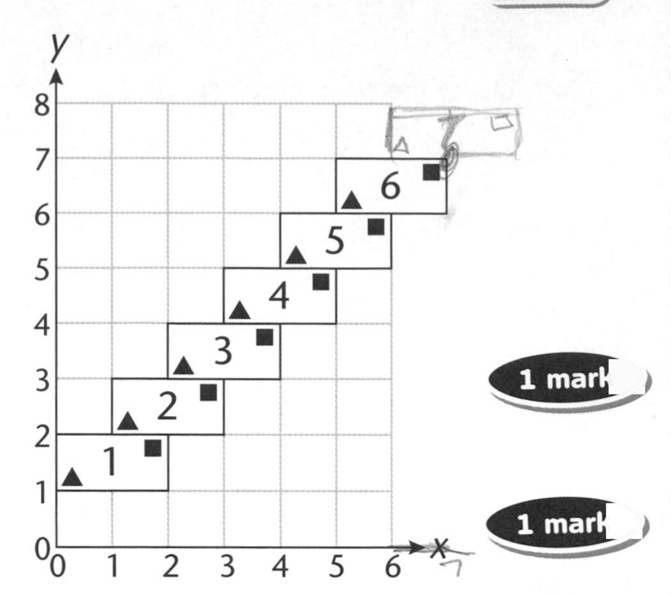

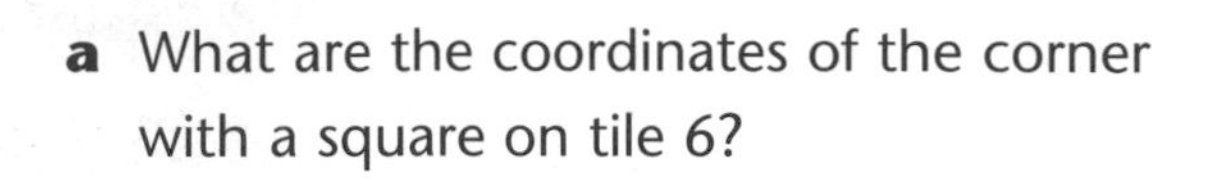

a What are the coordinates of the corner with a square on tile 6?

____

1 mark

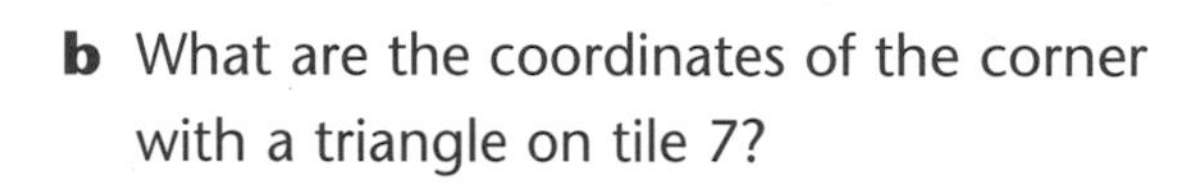

b What are the coordinates of the corner with a triangle on tile 7?

____

1 mark

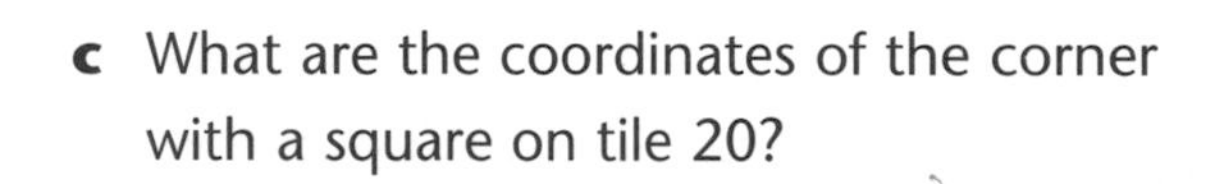

c What are the coordinates of the corner with a square on tile 20?

____

1 mark

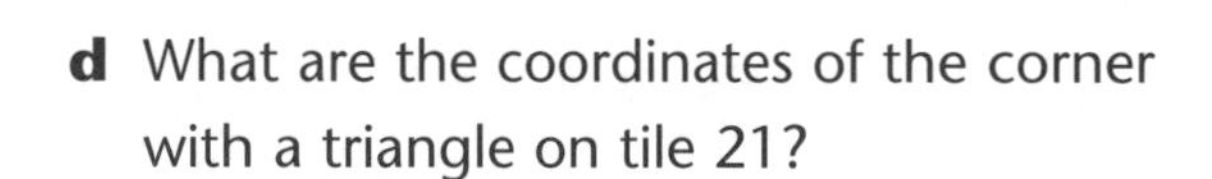

d What are the coordinates of the corner with a triangle on tile 21?

____

1 mark

2 The equation of a line is $y + x = 8$.

Which of the following points could lie on the line?

Tick the correct answer. (There could be more than one.)

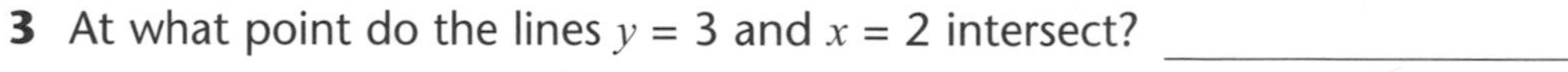

☐ (–2, –6) ☐ (0, 8) ☐ (–2, 10) ☐ (10, –2)

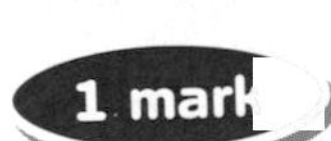

1 mark

3 At what point do the lines $y = 3$ and $x = 2$ intersect? _______________

1 mark

4 Which of these lines passes through the point (–3, 5)?

Tick the correct answer. (There could be more than one.)

☐ $x + y = 8$ ☐ $y = 5$

☐ $x + y = 2$ ☐ $x = -3$

1 mark

5 What are the equations of the lines a, b, c, d?

a _______________

b _______________

c _______________

d _______________

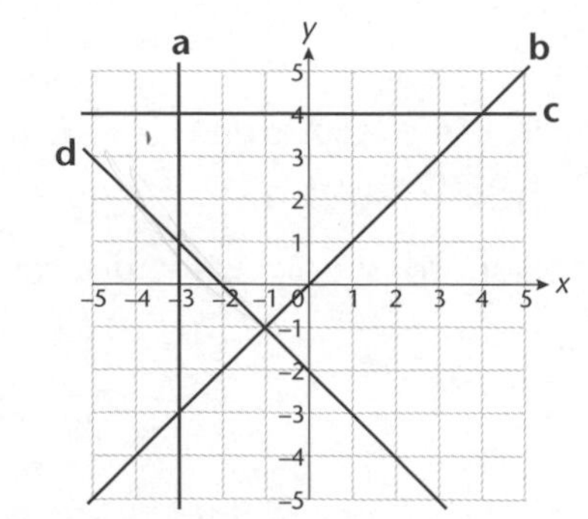

4 marks

levels 5-6

6 A, B and C are three corners of a triangle.

a What is the equation of the line

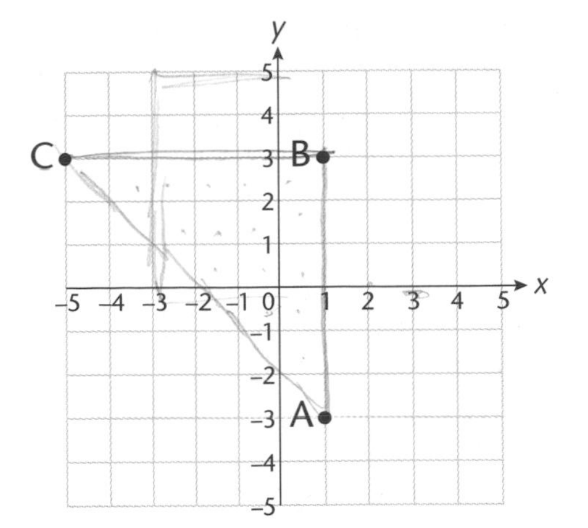

i AB ________________ 1 mark

ii BC ________________ 1 mark

iii AC ________________ 1 mark

b What is the area of the triangle ABC?

________________ 1 mark

7 What is the equation of the line shown?
Tick the correct answer.

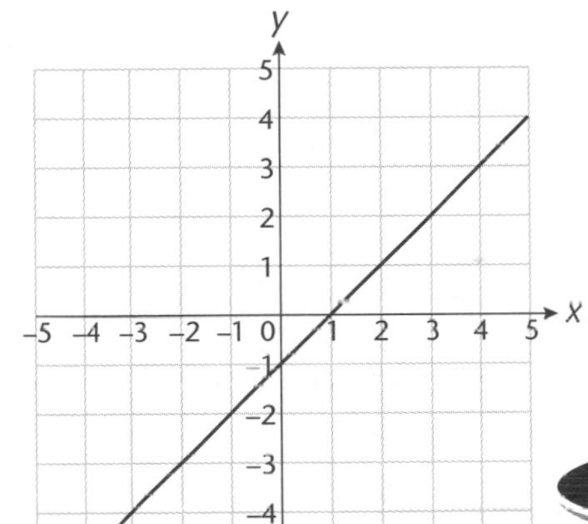

- [ ] $y = -1$
- [ ] $y = x - 1$
- [ ] $x = 1$
- [ ] $x + y = 1$

1 mark

8 A is the point (2, 5). B is the point (–3, –5).
Which of the following is the graph of the straight line through A and B?
Tick the correct answer.

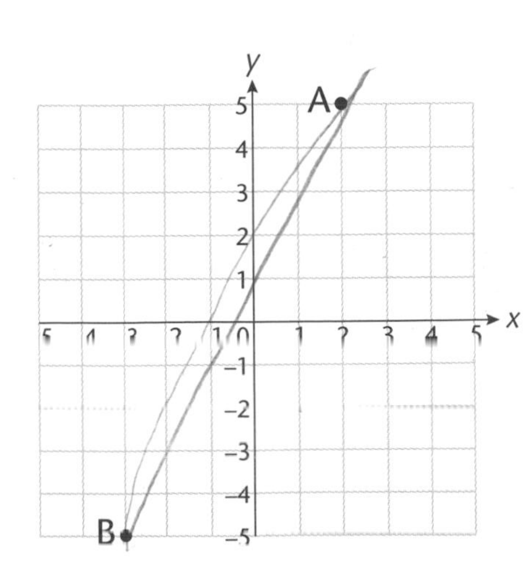

- [ ] $y = 2x - 1$
- [ ] $y = x + 3$
- [x] $y = 2x + 1$
- [ ] $y = x - 2$

1 mark

Graphs 2

1 The equations of four lines are

$y = 2x + 1$ $\quad y = 2x - 3$ $\quad y = 3x - 2$ $\quad y = 4x + 1$

a Which two lines are parallel?

____ **1 mark**

b Which two lines pass through the same point on the y-axis?

____ **1 mark**

2 Match the lines on the graph with the equations.

$y = x + 3$ matches line ________________

$y = 3x - 1$ matches line ________________

$x + y = -1$ matches line ________________

$y = -1$ matches line ________________

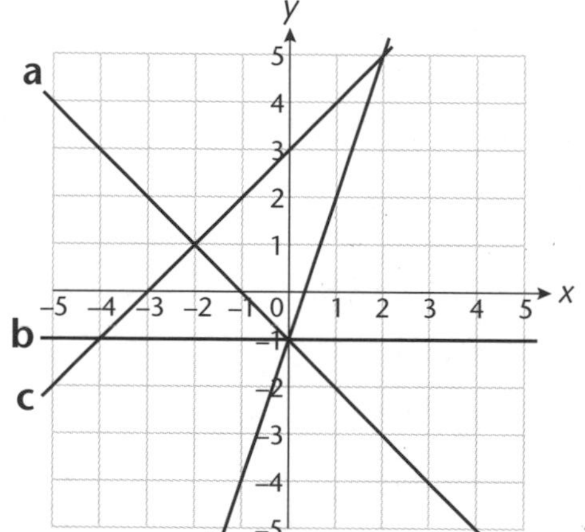

4 marks

3 The equation of a line is $y = 3x - 1$.
Which of the following points could lie on the line?
Tick the correct answer. (There could be more than one.)

☐ (2, 7) ☐ (3, 8) ☐ (–2, –7) ☐ (–3, 10)

1 mark

4 The solid line on the graph is $y = 2x + 5$.
What is the equation of the line parallel to this line that passes through (0, –2)?

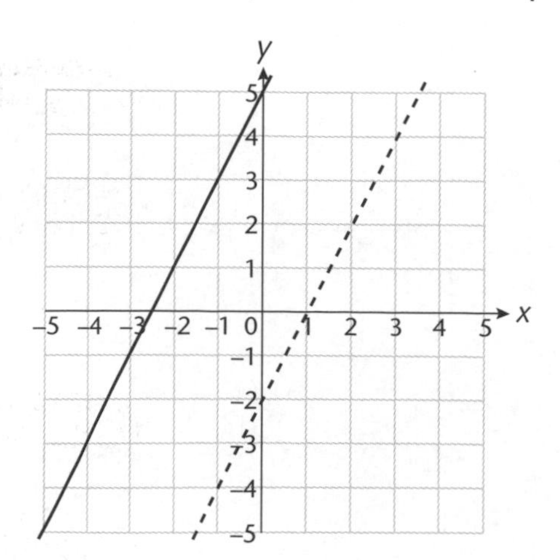

________________________ **1 mark**

levels 6-8

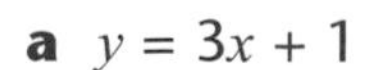

5 On the grid draw the graphs of

a $y = 3x + 1$

b $y = 2x - 3$

c $y = \frac{x}{2} + 3$

d $y = x - 4$

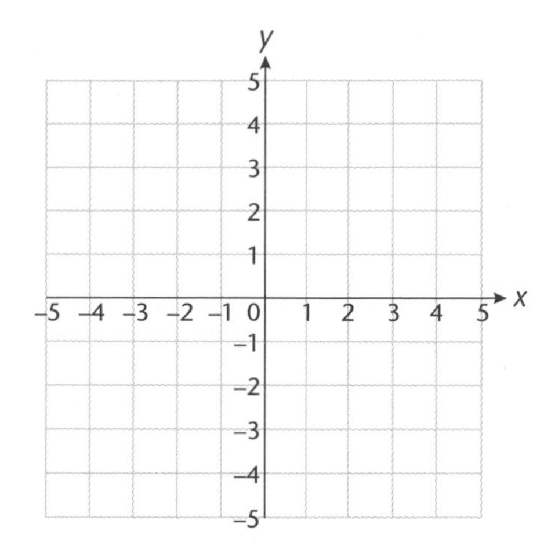

4 marks

6 Jenny does a parachute jump. Which graph shows the height against time?

A

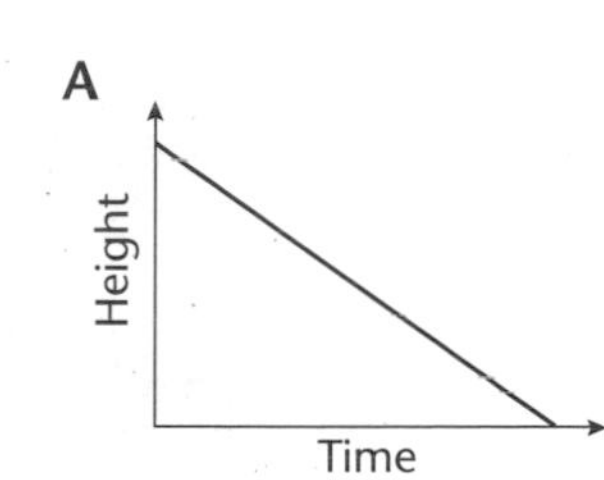

B

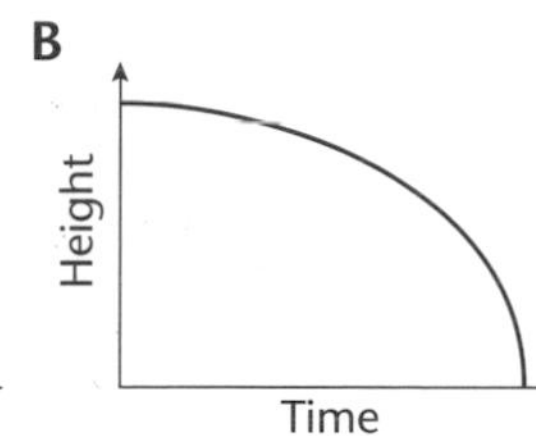

C

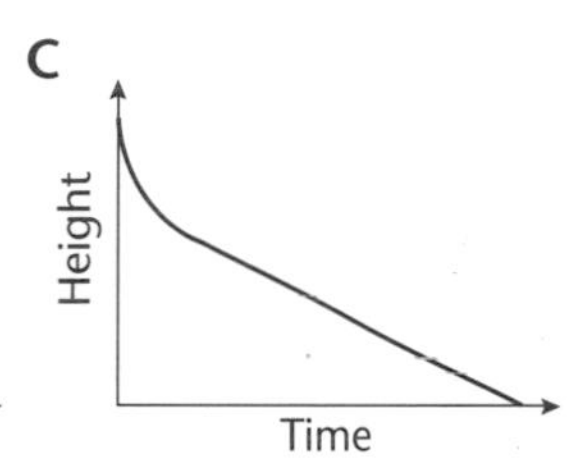

D

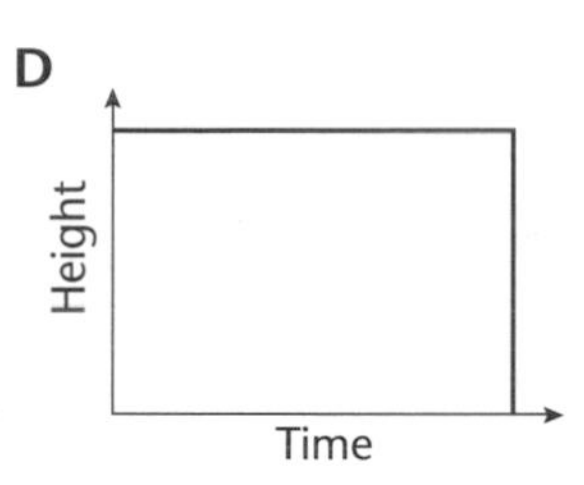

____

1 mark

7 Mr Stead does a 5 kilometre walk that takes him 50 minutes.

The distance–time graph shows his progress.

Distance (km): 0, 1, 2, 3, 4, 5
Time (minutes): 0, 10, 20, 30, 40, 50

a What is his average speed for the whole walk in kilometres per hour?

1 mark

b At one point Mr Stead took a rest.

i How many kilometres had he walked by the time he took a rest?

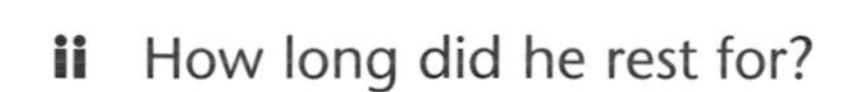

ii How long did he rest for?

____

1 mark

c Mr Stead ran the last kilometre.

i How long did he take to run the last kilometre?

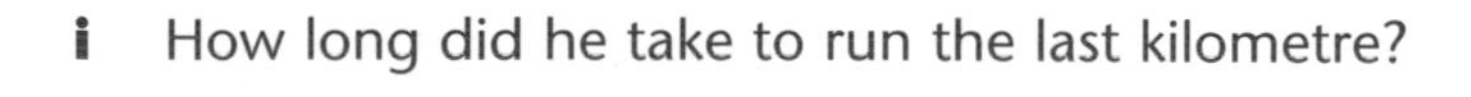

ii What was his average speed for the last kilometre?

____

1 mark

ALGEBRA Equations 1

levels 5-6

1 I think of a number. Add three to it. Divide the result by 5 and I get an answer of 6. What number did I think of?

____________ 1 mark

2 Solve the equation $2x - 7 = 20$

____________ 1 mark

3 Solve these equations.

a $\dfrac{x-2}{3} = 7$

____________ 1 mark

b $\dfrac{x}{3} - 2 = 7$

____________ 1 mark

4 Solve these equations.

a $3x - 8 = 7$

____________ 1 mark

b $\dfrac{x+3}{8} = 3$

____________ 1 mark

5 Solve these equations.

a $\dfrac{x+3}{5} = 6$

____________ 1 mark

b $\dfrac{x}{5} + 3 = 6$

____________ 1 mark

c $x + \dfrac{3}{5} = 6$

____________ 1 mark

d $5x - 3 = 6$

____________ 1 mark

6 Solve these equations.

a $\frac{x}{6} = \frac{7}{4}$

______________ **1 mark**

b $\frac{x}{4} = \frac{9}{2}$

______________ **1 mark**

c $\frac{5}{x} = \frac{2}{7}$

______________ **1 mark**

d $\frac{9}{2} = \frac{15}{x}$

______________ **1 mark**

7 Solve these equations.

a $\frac{x}{6} = \frac{7}{3}$

______________ **1 mark**

b $\frac{x + 3}{4} = \frac{1}{2}$

______________ **1 mark**

8 Complete the following statements to solve

$$\frac{2x - 5}{2} = \frac{x - 4}{4}$$

$(2x - 5) \times 4 = (x - 4) \times$ ______

$8x - 20 =$ ______

$8x -$ ______ $= -8 +$ ______

$6x = 12$

$6x \div 6 = 12 \div$ ______

$x =$ ______

2 marks

ALGEBRA Equations 2

levels 5-7

1 Solve the equation $x - 7 = 2x + 3$

2 Solve the equation $5x - 8 = 10 + 2x$

3 Solve these equations.

a $\dfrac{x + 2}{5} = 6 + x$

b $\dfrac{x}{5} + 5 = 3 + x$

c $x + \dfrac{3}{5} = 6 - x$

d $5x - 3 = 6 + x$

4 The following diagram shows a rectangle.

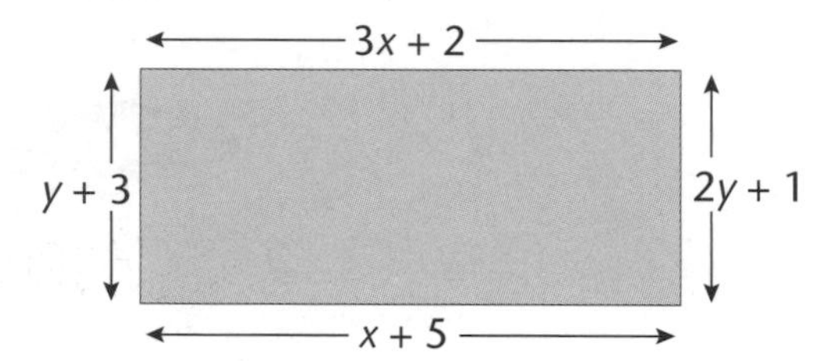

a Find the value of x.

1 mark

b Find the value of y.

5 The diagrams show some bricks. The bricks on the bottom row add up to the value or expression in the top row.

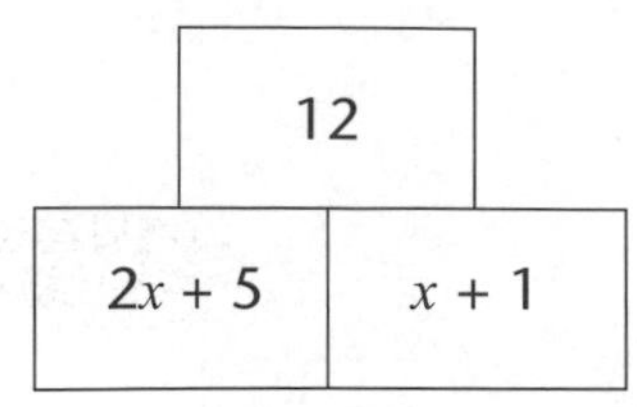

	$2y - 3$	
$3y + 1$		$y + 6$

a Find the value of x. 1 mark

b Find the value of y. 1 mark

levels 5-7

6 What inequalities are represented by the following?

a

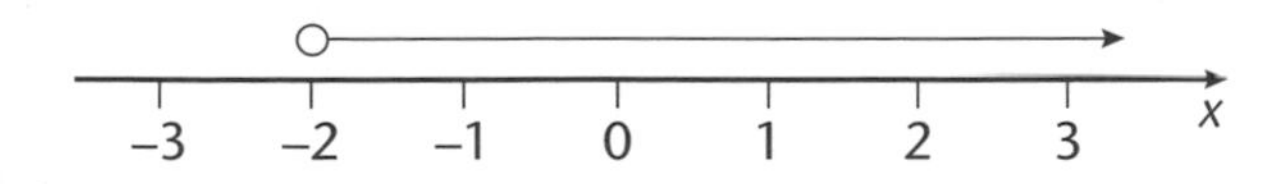

______________________ **1 mark**

b

−3 −2 −1 0 1 2 3 x

______________________ **1 mark**

7 Solve this inequality $2x - 7 < 21$

______________________ **1 mark**

8 Solve these inequalities.

a $3x + 8 \geq 20$

______________________ **1 mark**

b $3x - 8 > 19$

______________________ **1 mark**

9 Solve these inequalities.

a $4(x - 1) > 6$

______________________ **1 mark**

b $4(x + 3) \leq 12$

______________________ **1 mark**

ALGEBRA Trial and improvement

level 6

1 What is the value of 4^3? ______________

2 Estimate the value of 3.1^3. ______________

3 Estimate the value of $2.7^3 + 3 \times 2.7$. ______________

4 What is the value of $2^3 + 3 \times 2$? ______________

5 What is the value of $3.2^3 - 2 \times 3.2$?  ______________

1 mark

6 A rectangle has sides of x cm and $x + 3$ cm. It has an area of 40 cm^2.
Which of the following must be true?
Tick the correct answer. (There may be more than one.)

a ☐ $x(x + 3) = 40$

b ☐ The sides are 2 cm and 20 cm

c ☐ The sides are 5 cm and 8 cm

d ☐ The perimeter is 26 cm

7 Complete the table to find a solution to the equation: 
$x^3 = 100$
Give your answer to 1 decimal place.

x	x^3	Comment
4	64	Too low
5	125	Too high

x = ______________

level 6

8 Complete the table to find a solution to the equation: 

$x^3 + 3x = 20$

Give your answer to 1 decimal place.

x	$x^3 + 3x$	Comment
2	14	Too low

$x =$ ____________

3 marks

9 A rectangle has sides of x cm and $x + 2$ cm. It has an area of 16.64 cm^2. 

a Explain why $x^2 + 2x = 16.64$

____

____

1 mark

b Complete the table to find the value of x.

x	$x^2 + 2x$	Comment
2	8	Too low

$x =$ ____________

2 marks

Simultaneous equations

level 8

1 Which of the following equations do the values $x = 2$ and $y = 3$ satisfy?
Tick the correct answer. (There may be more than one.)

Equation 1 $y = 2x + 1$ ☐ Equation 2 $y = 2x - 1$ ☐

Equation 3 $2y + x = 8$ ☐ Equation 4 $2y - x = 1$ ☐

1 mark

2 **a** Show that if $y = 2x + 3$ and $2y + x = 11$ then $5x + 6 = 11$

b Solve the equation $5x + 6 = 11$

c Find the values of x and y that satisfy these simultaneous equations.
$y = 2x + 3$ and $2y + x = 11$

$x =$ ______________

$y =$ ______________

3 Both of these rectangles have a perimeter of 20 cm.

a Show that $x + y = 4$
and $2x + 3y = 10$

$3y + 1$

$2x - 1$

$3y - 2$

$3x$

2 marks

b Solve the simultaneous equations to find the values of x and y.

$x =$ ______________

$y =$ ______________

4 Solve these simultaneous equations.

$$y = 2x - 1$$
$$4y = 6x + 2$$

$x =$ ______________

$y =$ ______________

5 One cup of tea and a sticky bun cost £1.20.
Three teas and two sticky buns cost £3.10.

a What is the cost of one tea? __________ **1 mark**

b What is the cost of a sticky bun? __________ **1 mark**

6 The diagram shows an isosceles triangle.
The perimeter of the triangle is 20 cm.

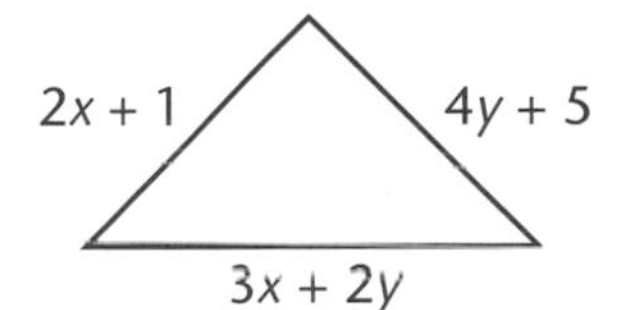

a Show that $x - 2y = 2$

2 marks

b Show that $5x + 6y = 14$

2 marks

c Solve the simultaneous equations $x - 2y = 2$ and $5x + 6y = 14$.

$x =$ __________

$y =$ __________ **2 marks**

7 Two families visit the cinema.
The Watsons buy 2 adult tickets and 2 children's tickets for £14.00.
The Cricks buy 3 adult tickets and 1 child ticket for £16.00.
Let x be the cost of an adult ticket and y be the cost of a child ticket.

a Set up two simultaneous equations using the information above.

____________________ **2 marks**

b Solve the simultaneous equations to find the cost of an adult and a child ticket.

Adult $x =$ __________

Child $y =$ __________ **2 marks**

Bearings

level 6

1 Fill in the missing bearings on this diagram.

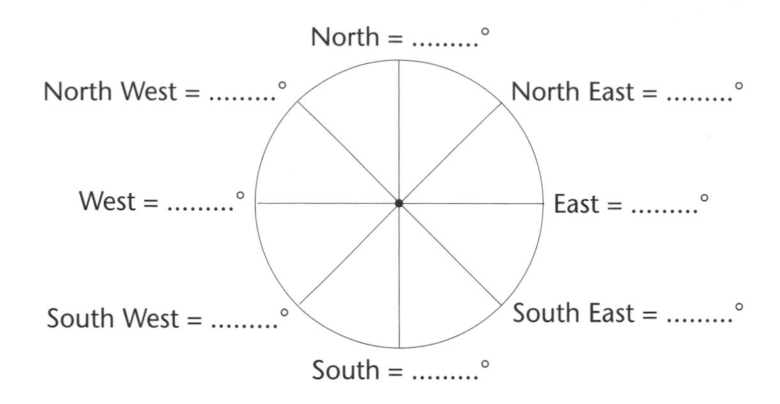

8 marks

2 From the point O draw the positions of the points A, B, C, and D that lie on the circle and have bearings of

A 080°; B 120°; C 220°; D 300°

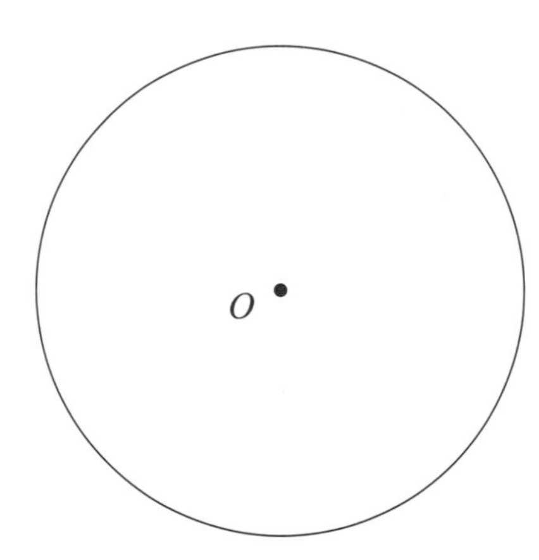

4 marks

3 Estimate the bearings of each of these lines.

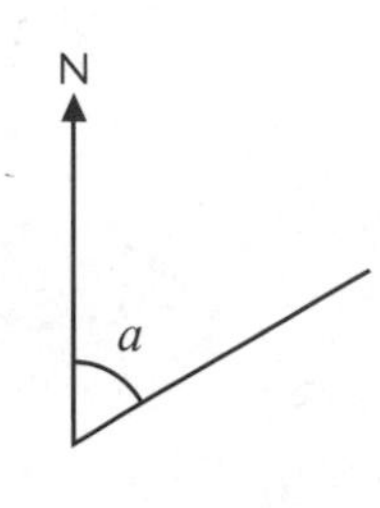

N

b

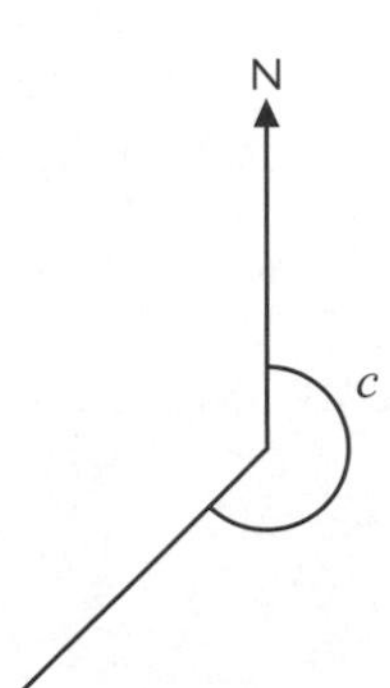

a = ________ b = ________ c = ________

3 marks

level 6

4 Measure **a** the actual distance and **b** the bearing of *A, B* and C from *O* in the diagram below.

Scale 1 cm : 2 km

C

B

6 marks

5 From the point *O* draw the following points.

a *A* which is 5 cm on a bearing of 065° from *O*

b *B* which is 4 cm on a bearing of 155° from *O*

c *C* which is 6 cm on a bearing of 265° from *O*

N

O

6 marks

SHAPE, SPACE AND MEASURES

Angle facts

levels **5-6**

1 *PQR* is an isosceles triangle. Calculate the value of angle g.

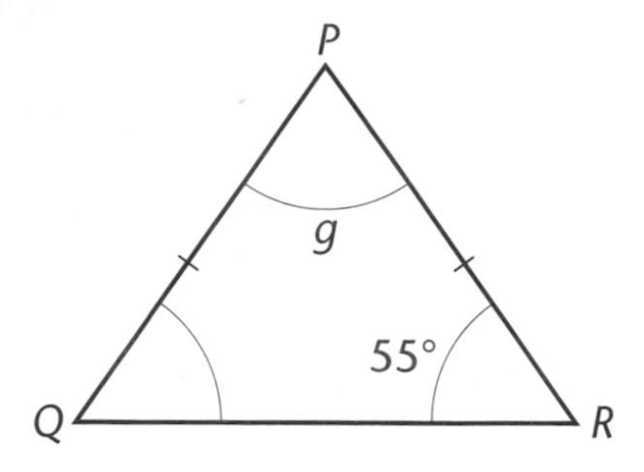

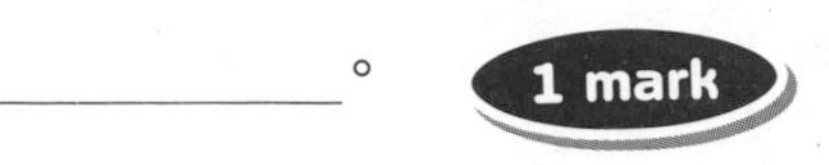

______ °　**1 mark**

2 Calculate the size of angle x.

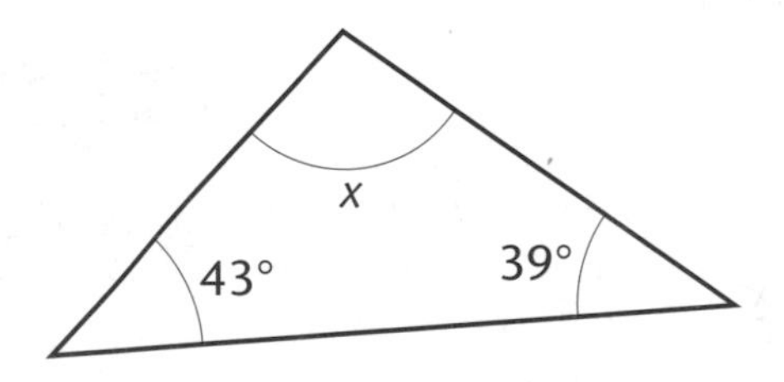

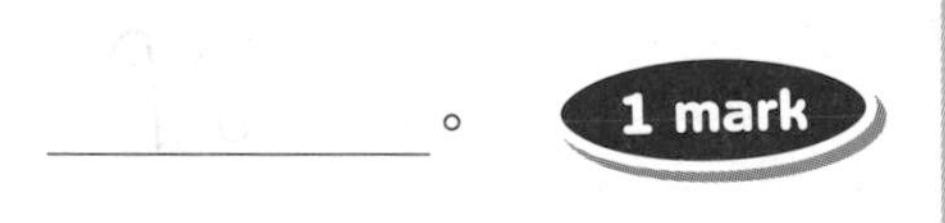

______ °　**1 mark**

3 Calculate the size of angle y.

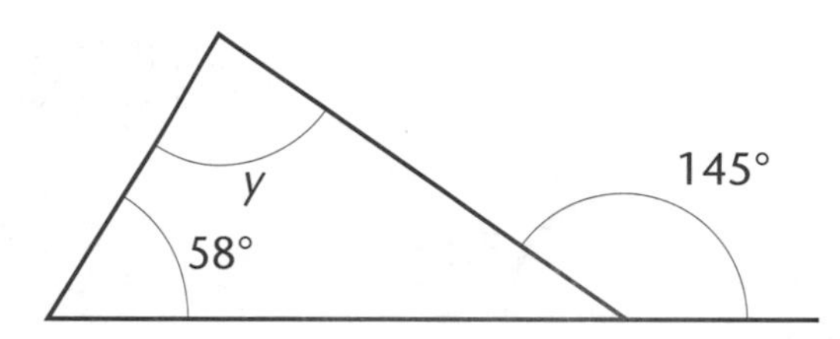

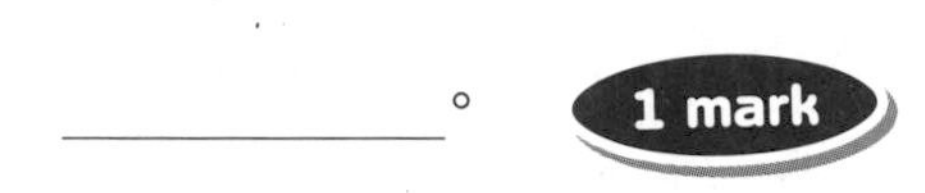

______ °　**1 mark**

4 Calculate the size of angle z.

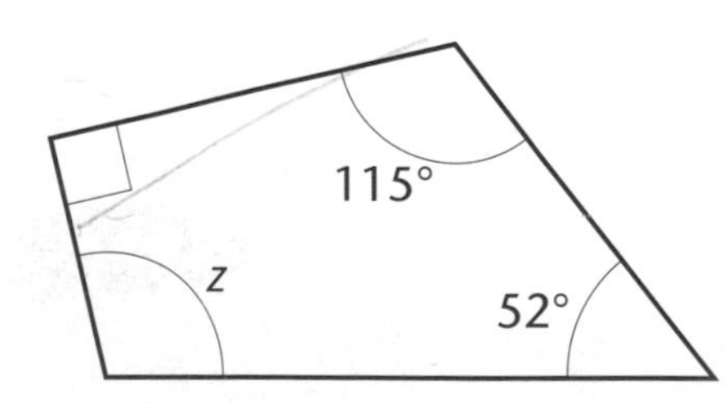

______ °　**1 mark**

5 *ABC* is a right-angled triangle. *ACD* is an isosceles triangle.
Angle *BAC* = 35°. Angle *CAD* = 40°.

a Find angle *CDA*.

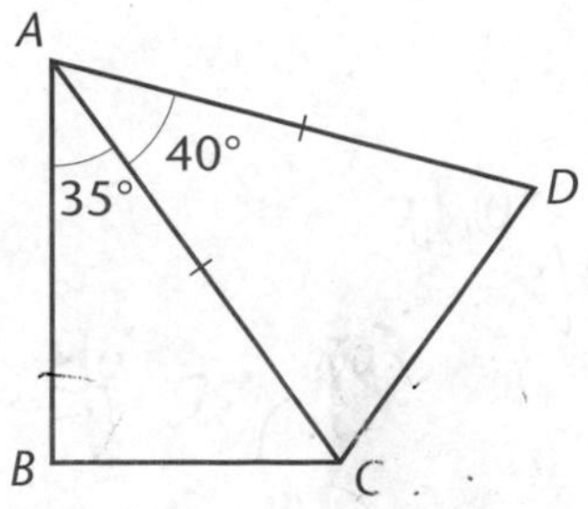

______ °　**1 mark**

b Find angle *BCD*.

______ °　**1 mark**

6 Using the diagram explain why

$z = x + y$

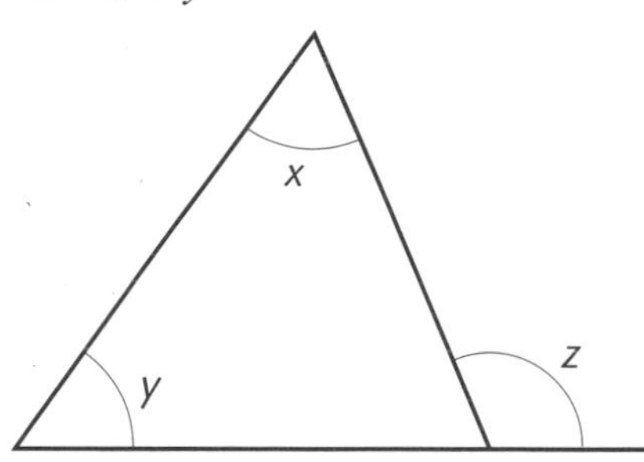

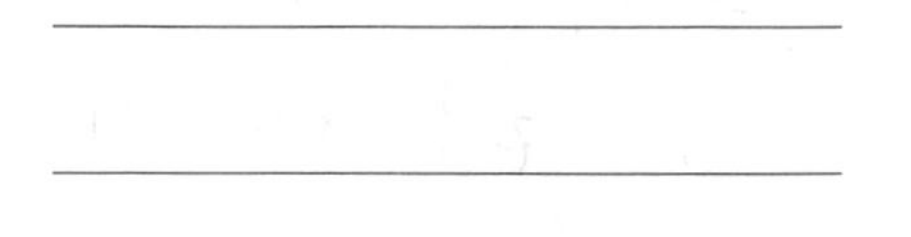

2 marks

7 The diagram shows a rectangle *ABEF* inside a regular octagon *ABCDEFGH*.

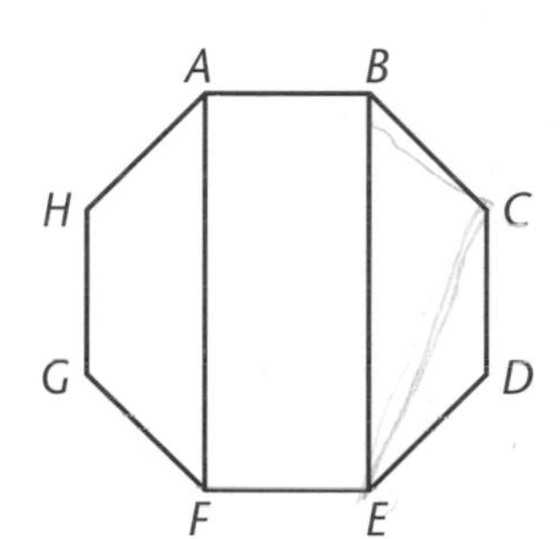

a What shape is the quadrilateral *BCDE*?

b What is value of the angle *EBC*?

______ °

8 The diagram shows an equilateral triangle *ABC* and a regular pentagon *BCDEF*.

Work out the value of the angle *FBA*.

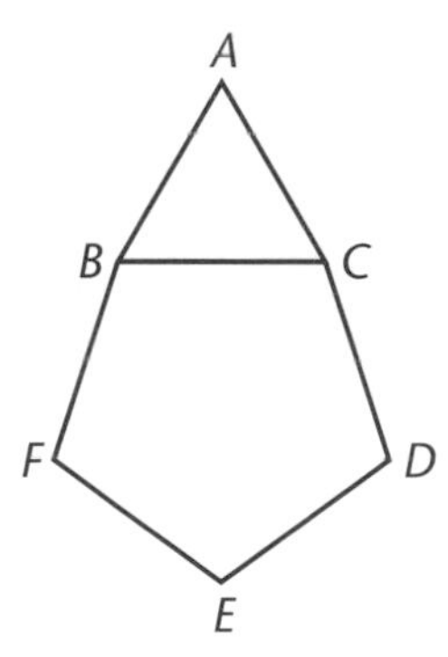

______ °

1 mark

9 This star is constructed on a regular pentagon.

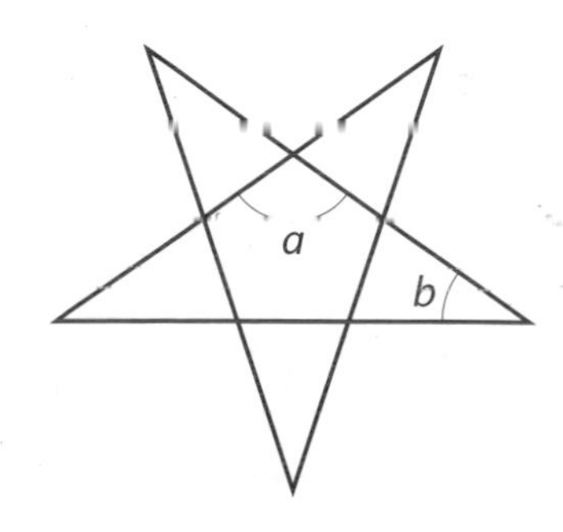

a Write down the value of angle a.

______

1 mark

b Write down the value of angle b.

______

Angles in parallel lines and polygons

levels 4-6

1 Write down the values of angles a, b and c.

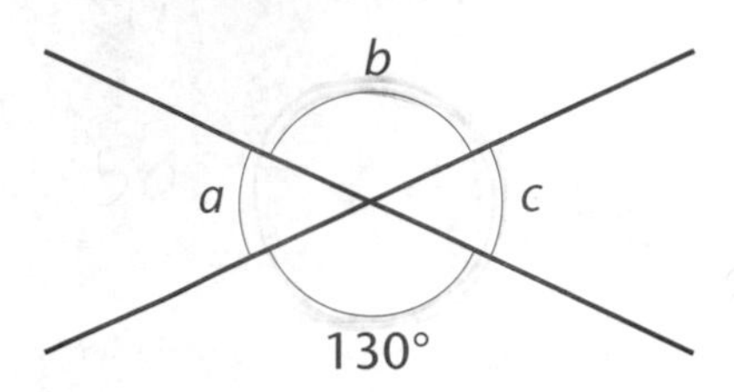

a = ________ b = ________ c = ________

2 Write down the value of angle d. Give a reason for your answer.

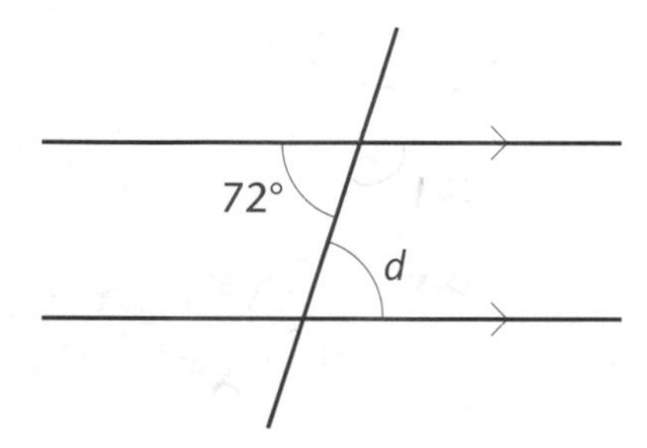

d = ________ because ________________________

3 Write down the value of angle e. Give a reason for your answer.

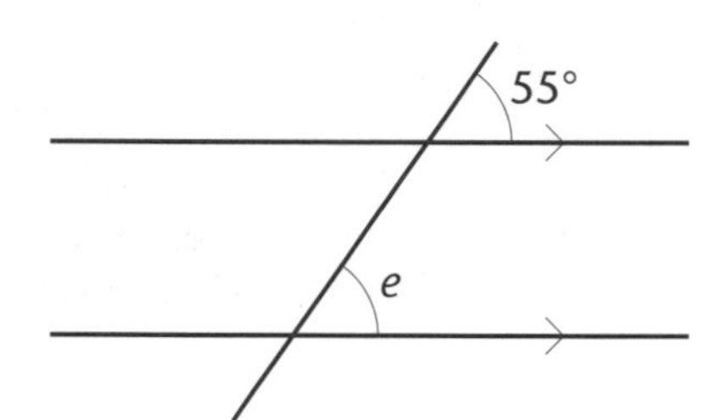

e = ________ because ________________________

4 Write down the value of angle f. Give a reason for your answer.

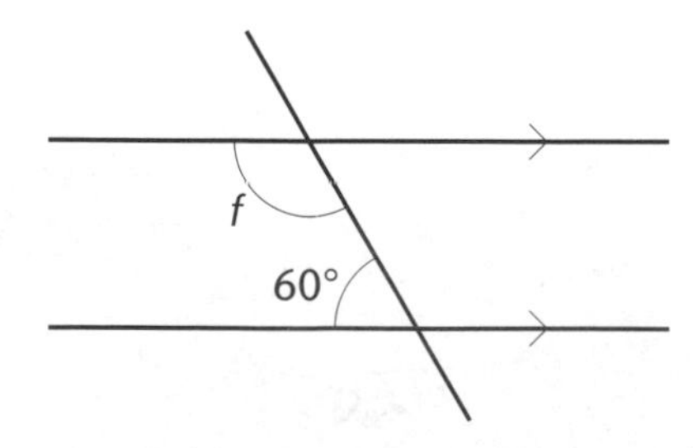

f = ________ because ________________________

5 Write down the value of angles g and h. Give reasons for your answers.

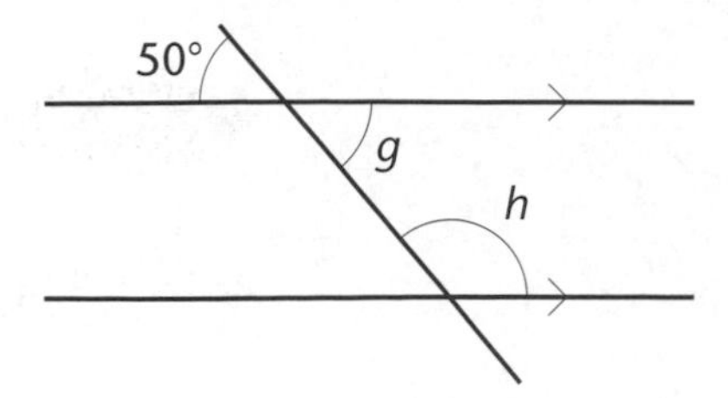

g = ________ because ________________________

h = ________ because ________________________

levels 4-6

6 What is the sum of the interior angles of a pentagon?

________ **1 mark**

7 Which of the following statements is true for a regular hexagon?
Tick the correct answer. (There may be more than one.)

☐ Each interior angle is 60° ☐ Each interior angle is 120°

☐ Each exterior angle is 60° ☐ Each exterior angle is 120°

1 mark

8 *ABCDE* is a regular pentagon.
Work out the value of angles x, y and z.

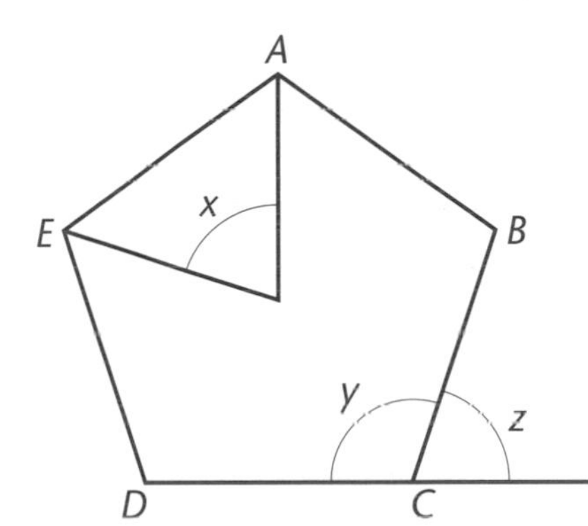

$x =$ ________ $y =$ ________ $z =$ ________

3 marks

9 The diagram shows four regular octagons, A, B, C and D.
Explain why shape S is a square.

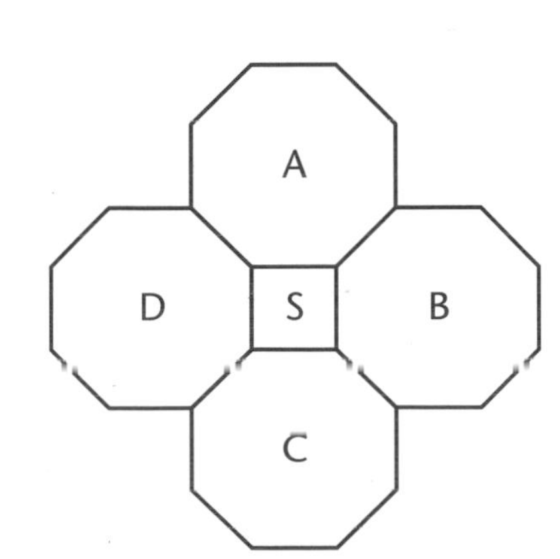

____

____

____

2 marks

Reflections and rotations

level 6

1 The grid shows a shaded triangle that has been rotated **clockwise** to the triangle A.

a What is the angle of rotation?

1 mark

b Mark the centre of rotation on the grid.

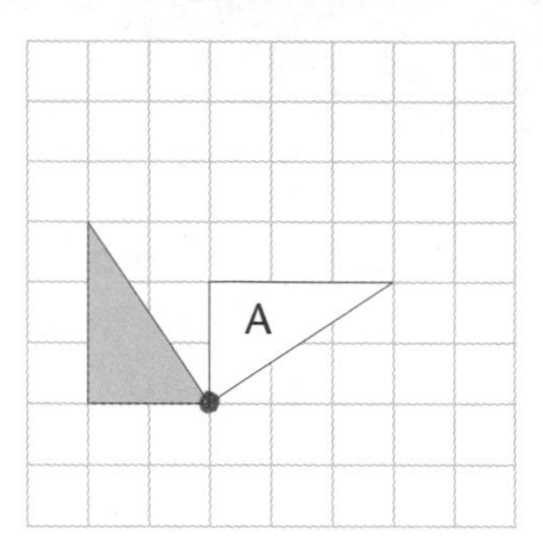

1 mark

2 The grid shows triangle A that has been reflected onto triangle B.

Draw the mirror line on the grid.

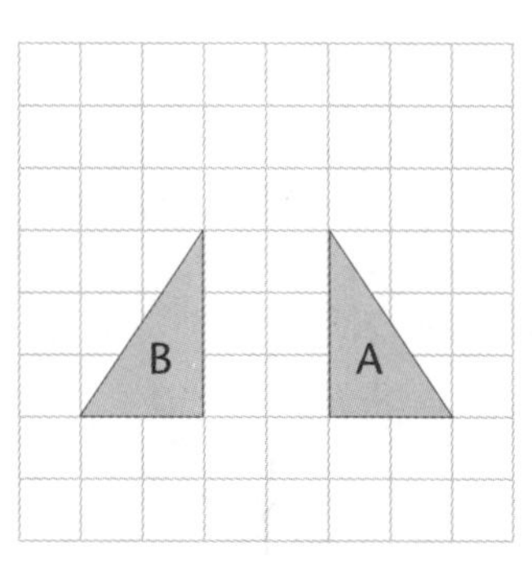

1 mark

3 The grid shows two transformations of the shaded triangle.

a Describe fully the **single** transformation that takes the shaded triangle to triangle A.

____

____

3 marks

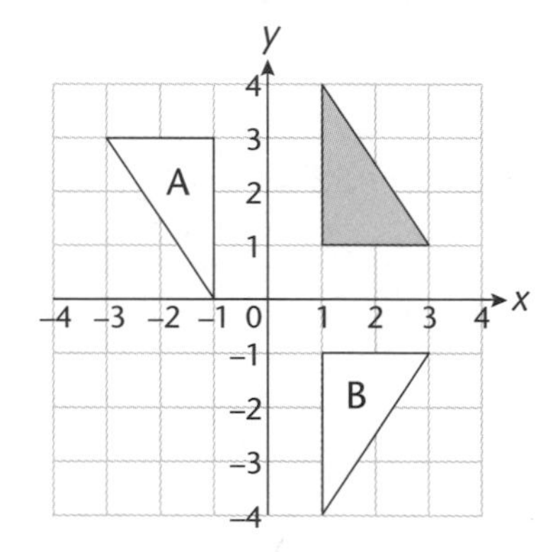

b Describe fully the **single** transformation that takes the shaded triangle to triangle B.

____

____

2 marks

4 Draw the reflection of the shape shown in the mirror line $x = -1$.

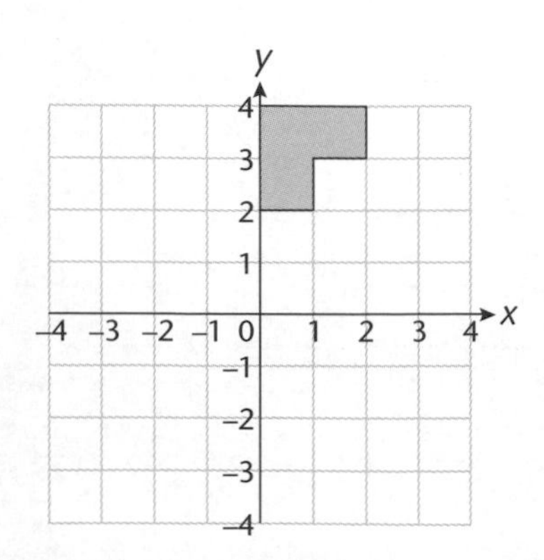

1 mark

level 6

5 Reflect the given shape in the line $y = x$.

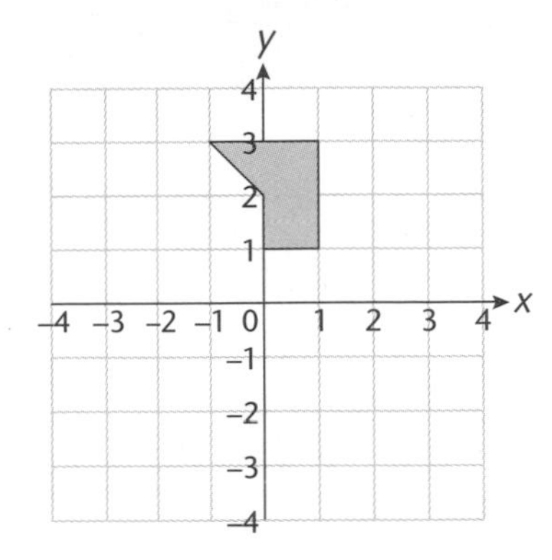

1 mark

6 Rotate the shaded triangle anticlockwise by 90° about (–1, 0).

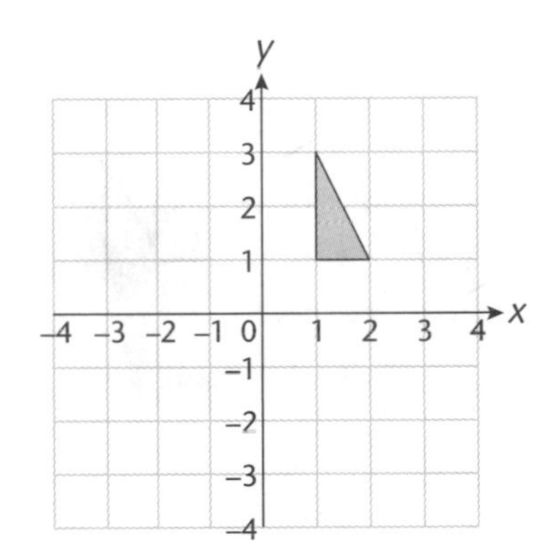

1 mark

7 Describe fully the transformation that takes the shaded triangle to

a triangle A

____

____

2 marks

b triangle B

____

____

2 marks

c triangle C

____

____

____

3 marks

Enlargements

level 6

1 Shapes A and B are enlargements of the shaded shape.

What is the scale factor of each enlargement?

A ____________________

B ____________________

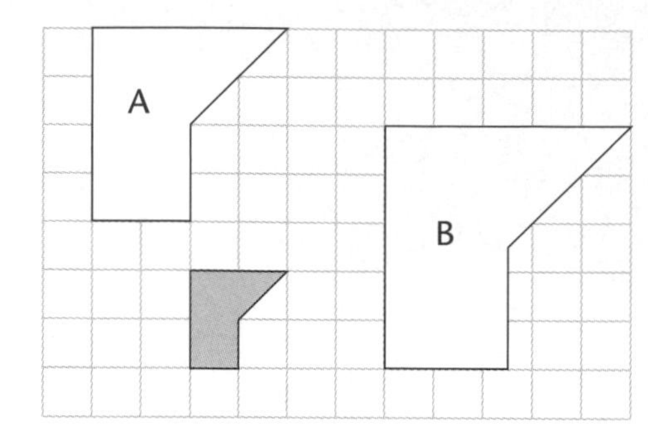

2 marks

2 The white shape is an enlargement of the shaded shape with scale factor 2.

One 'ray' joining points is shown.

Draw the other rays to find the centre of enlargement.

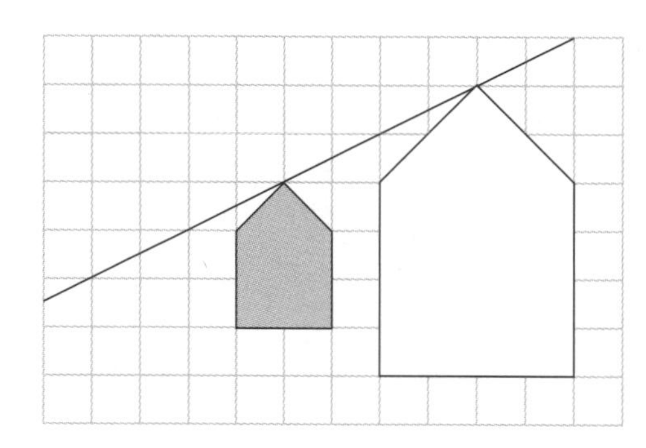

1 mark

3 On the grid, draw an enlargement of the triangle with a scale factor of 2.

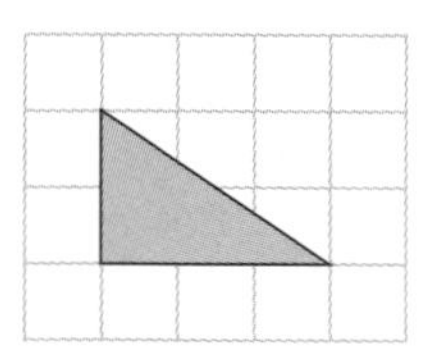

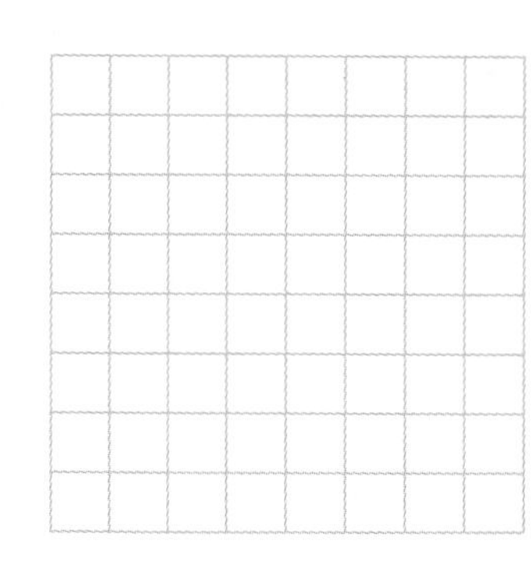

1 mark

4 On the grid, draw an enlargement of the shape with a scale factor of 3.

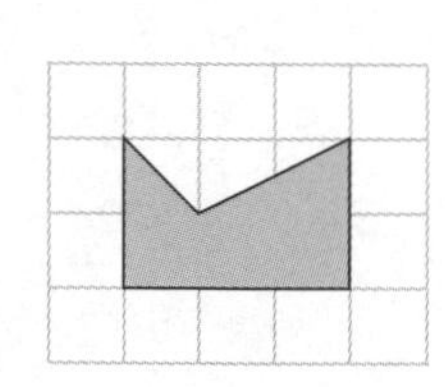

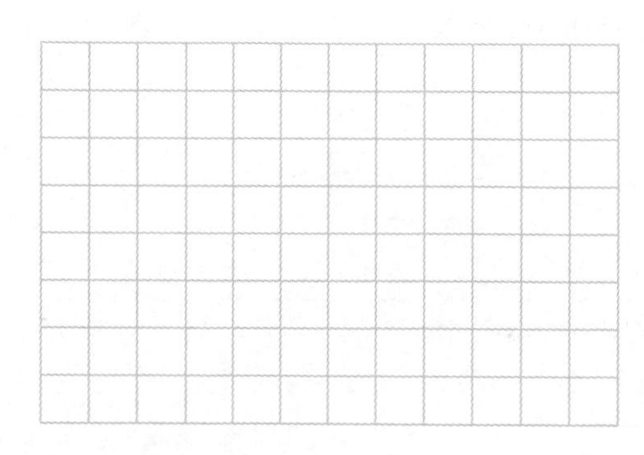

1 mark

level 6

5 The triangle ABC is enlarged to a triangle $A'B'C'$ by a scale factor of 2 about the origin.

Write down the coordinates of the points A', B' and C'.

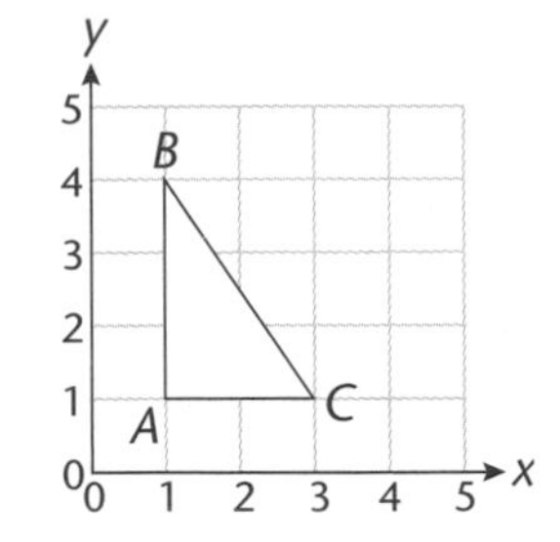

A' (______ , ______)

B' (______ , ______)

C' (______ , ______)

3 marks

6 The triangle $A'B'C'$ has been enlarged from a triangle ABC by a scale factor of 3 about the origin.

Write down the coordinates of the points A, B and C.

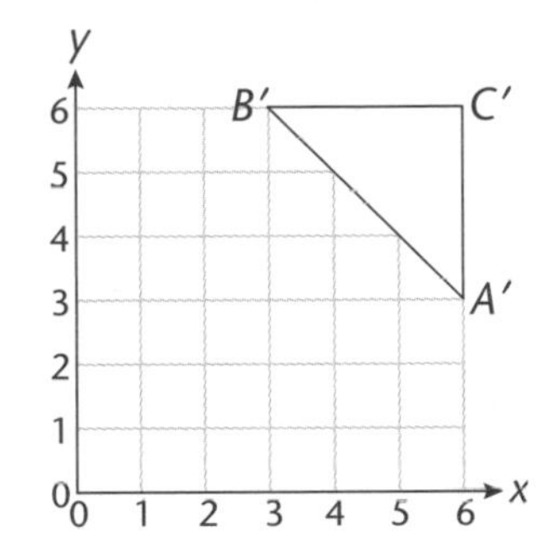

A (______ , ______)

B (______ , ______)

C (______ , ______)

3 marks

7 The shaded triangle has been transformed to triangles A, B and C.

Match the triangle to the transformation described below.

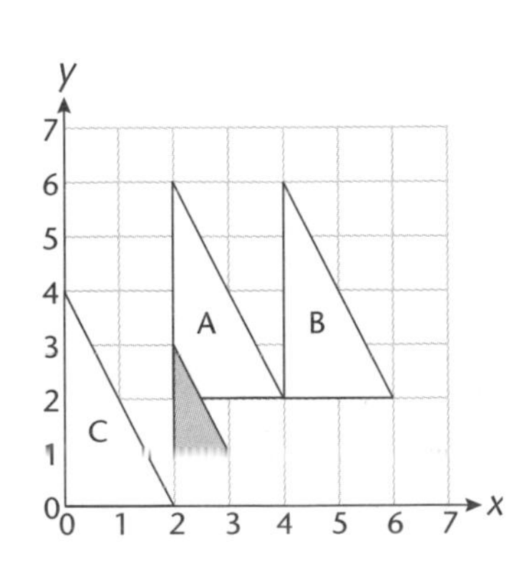

a Enlargement scale factor 2 about (0, 0) is

____

b Enlargement scale factor 2 about (2, 0) is

____

c Enlargement scale factor 2 about (4, 2) is

____

3 marks

3-D shapes

levels 5-7

1 Look at the net.

What is the name of the shape that will be formed by this net?

____

1 mark

2 Which of the following are nets for a cube?

a

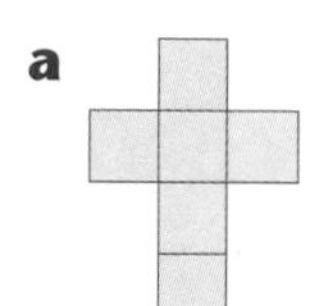

b

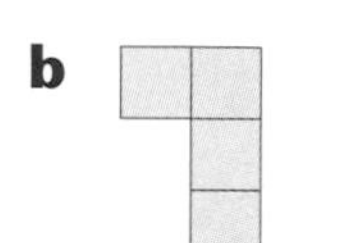

c

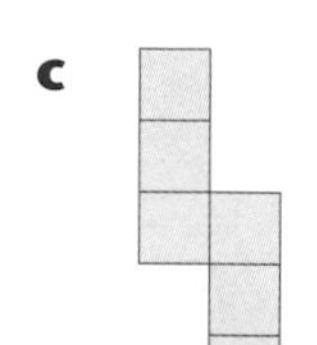

d

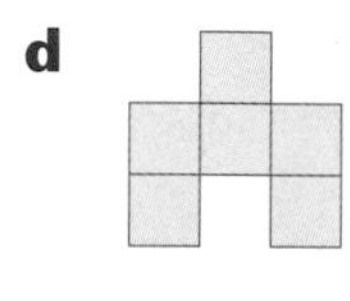

____

1 mark

3 This is the plan and elevation for a solid.

What is the name of this solid?

____

PLAN

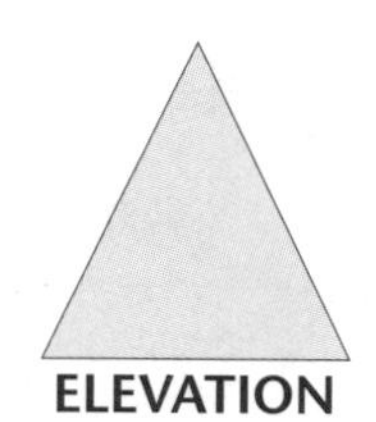

ELEVATION

1 mark

4 For this solid, draw

a the plan

b the elevation from X

c the elevation from Y

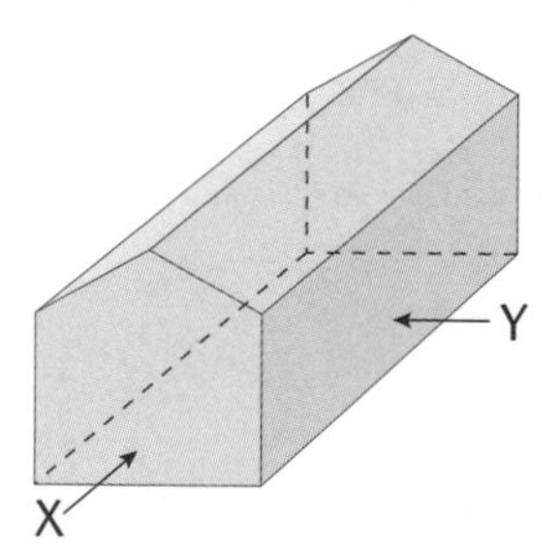

3 marks

5 How many planes of symmetry do the following shapes have?

a

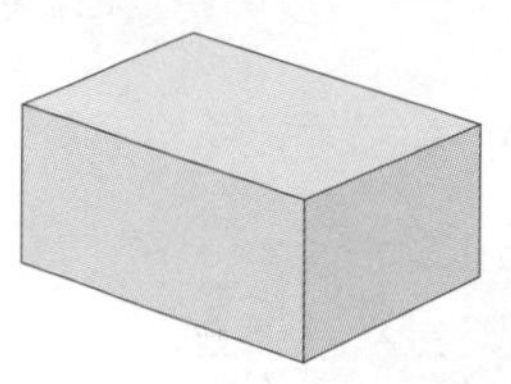

b

c

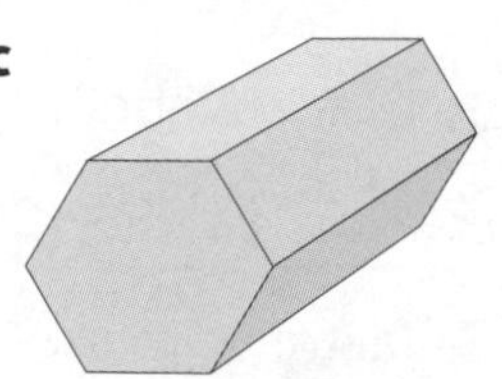

____________________ ____________________ ____________________

3 marks

levels 5-7

6 The shape shown is made from six centimetre cubes.

Draw **a** the plan

b the elevation from A

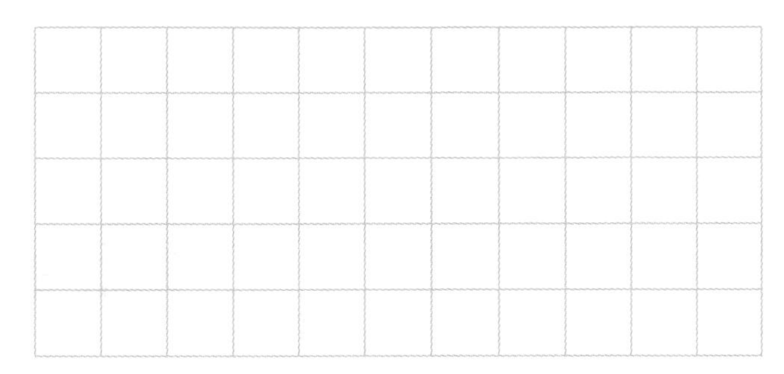

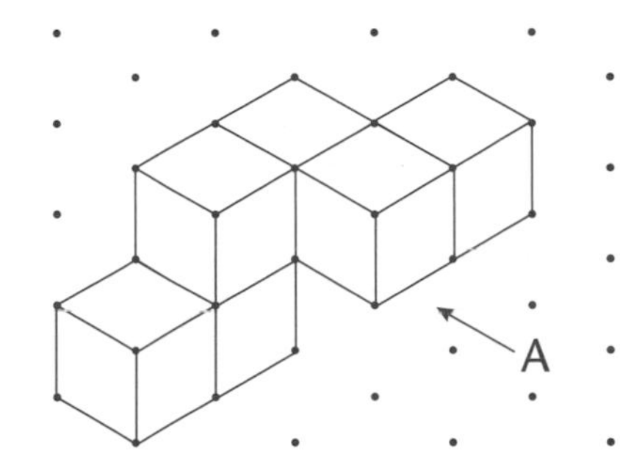

2 marks

7 A shape is made from four centimetre cubes.

The plan and two side elevations are shown below.

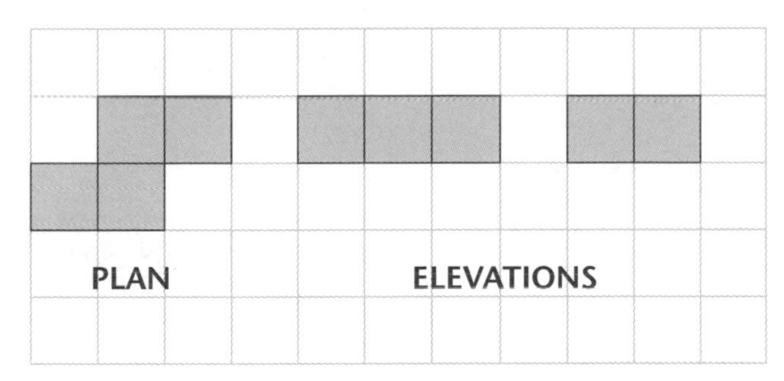

2 marks

Draw an isometric view of the shape.

8 A cylinder has a height of h and a radius of r.

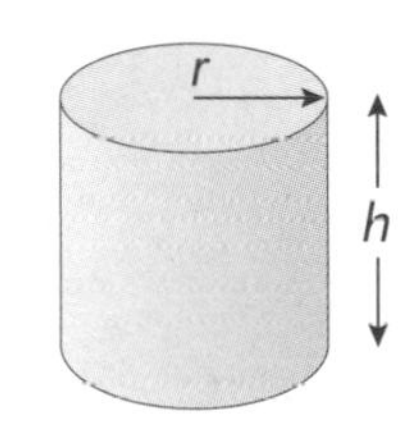

a Which of the following is the formula for the surface area of the cylinder?

i $\pi r^2 + rh$ **ii** $2\pi r^2 + 2\pi rh$ **iii** $\pi r^2 h$

____

b Work out **i** the volume and **ii** the surface area of a cylinder with height 10 cm and radius 6 cm. Give your answers to 3 significant figures.

i Volume ______________ cm^3

ii Surface area ______________ cm^2

9 Work out the surface area of a cuboid with length 12 cm, width 6 cm and height 8 cm.

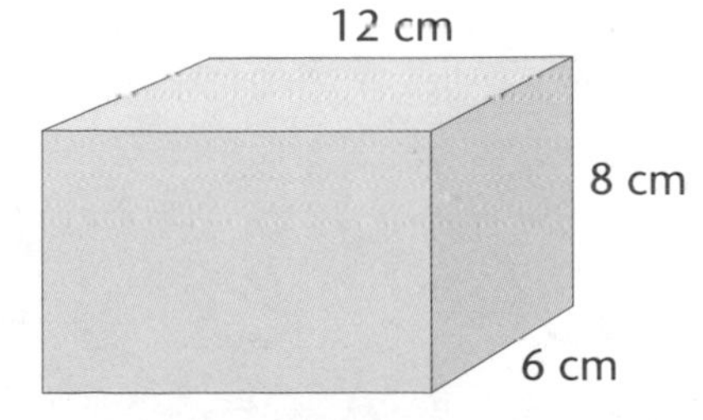

______________________ cm^2

2 marks

SHAPE, SPACE AND MEASURES

Perimeter and area

levels 5-6

1 What is **a** the perimeter and **b** the area of this right-angled triangle?

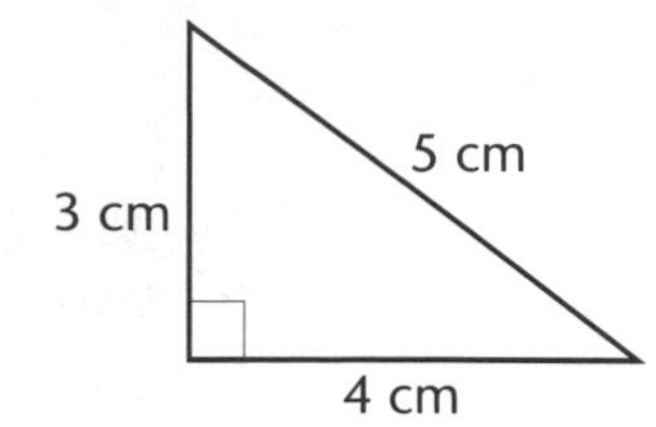

a perimeter ____________________

b area ____________________

2 Work out **a** the perimeter and **b** the area of this isosceles triangle.
Remember to include the units in your answer.

a perimeter ____________________

b area ____________________

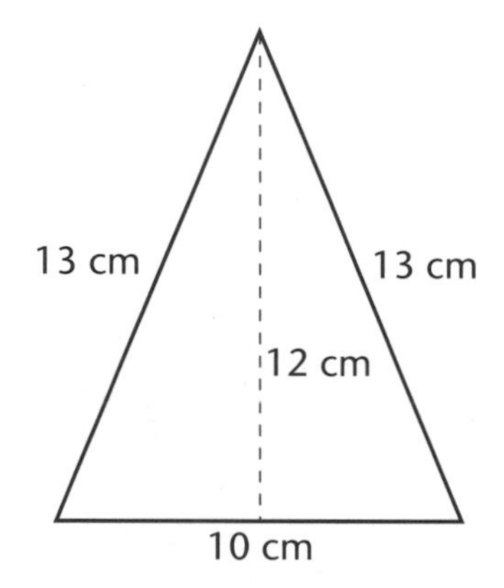

3 A, B, C and D are triangles drawn on a centimetre grid.
What are the areas of triangles A, B, C and D?

A ____________________ cm^2

B ____________________ cm^2

C ____________________ cm^2

D ____________________ cm^2

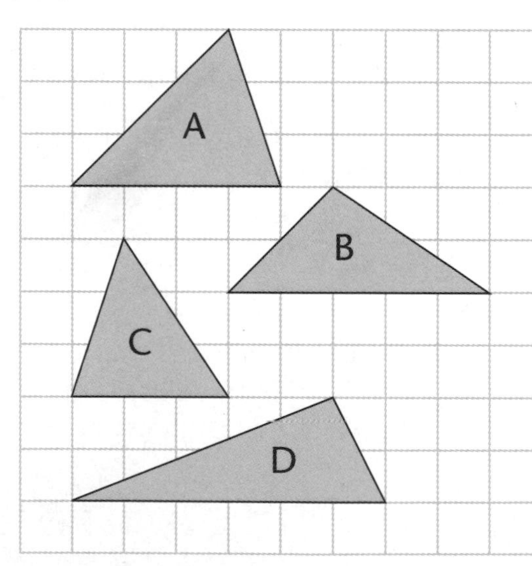

4 What is the area of this parallelogram?

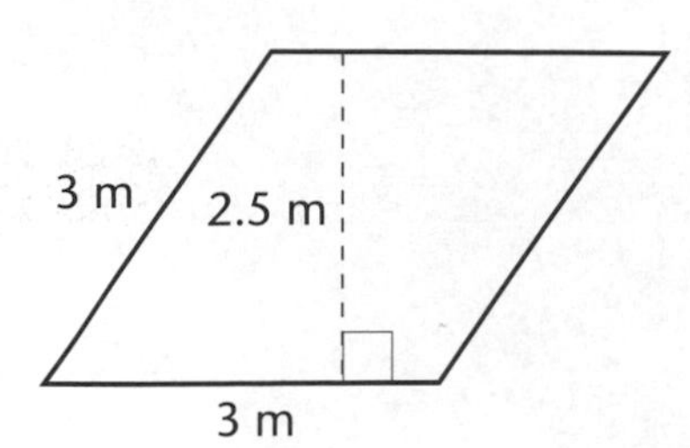

____

2 marks

5 What is the area of this trapezium?

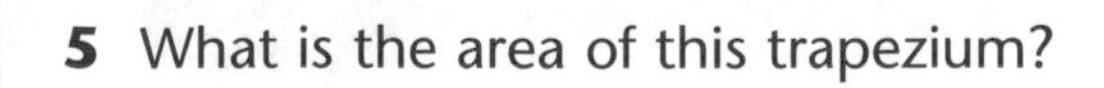
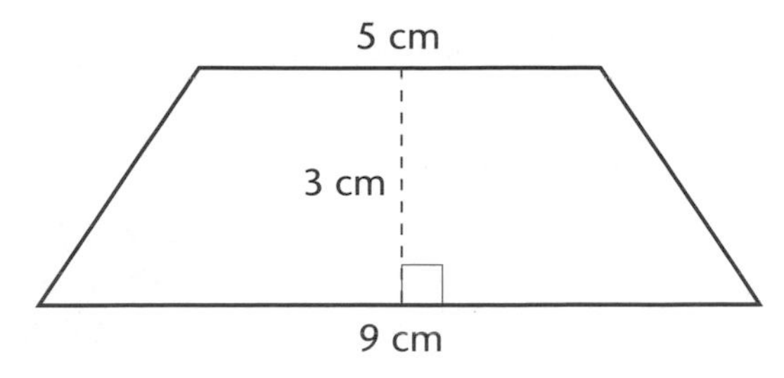

_______________ 2 marks

6 All of the following shapes have the same area. 

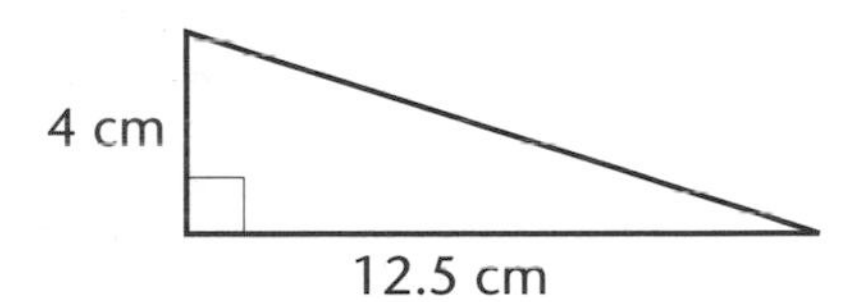

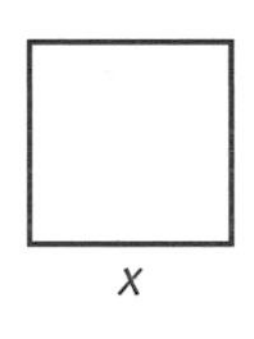

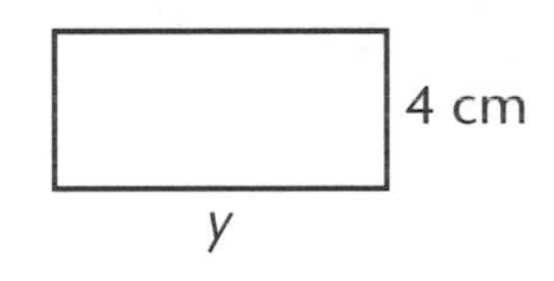

a Find the value of x. _______________ 1 mark

b Find the value of y. _______________ 1 mark

7 Find the area of this shape.

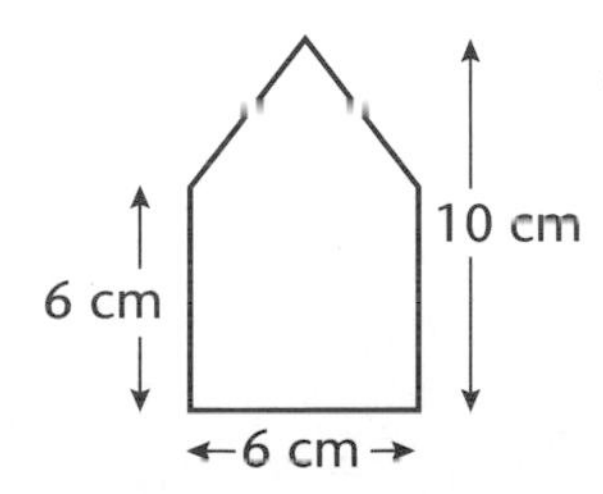

_______________ 2 marks

Circumference and area of a circle

level 6

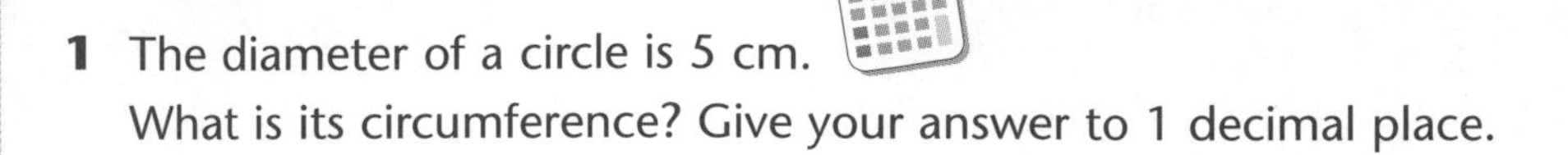

1 The diameter of a circle is 5 cm.
What is its circumference? Give your answer to 1 decimal place.

______________________ cm **1 mark**

2 The radius of a circle is 4 m.
What is its circumference? Give your answer to 1 decimal place.

______________________ m **1 mark**

3 The circumference of a circle is 25 cm.
What is its diameter? Give your answer to the nearest centimetre.

______________________ cm **1 mark**

4 A tin of beans has a diameter of 7.5 cm.
The label around the tin has an overlap of 1 cm.
What is the length of the label?
Give your answer to 1 decimal place.
Remember to include the units in your answer.

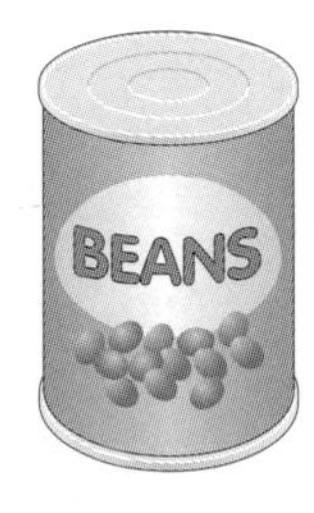

______________________ **2 marks**

5 What is the perimeter of this semicircle?
Give your answer to 1 decimal place.

10 cm

______________________ cm **1 mark**

6 The radius of a circle is 3 cm.
What is its area? Give your answer to 1 decimal place.

______________________ cm^2 **1 mark**

7 The diameter of a circle is 5 cm. 
What is its area?
Remember to include the units in your answer.

____________________ **2 marks**

8 A circle has a diameter of 18 cm.
What is its area?
Give your answer as a multiple of π.

____________________ cm² **1 mark**

9 What is the area of this quadrant?
Give your answer to 1 decimal place.

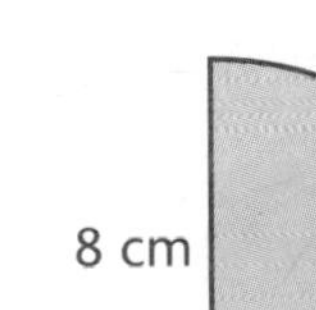

____________________ cm² **1 mark**

10 What is the area of the shaded part of the diagram? 
Give your answer to 1 decimal place.

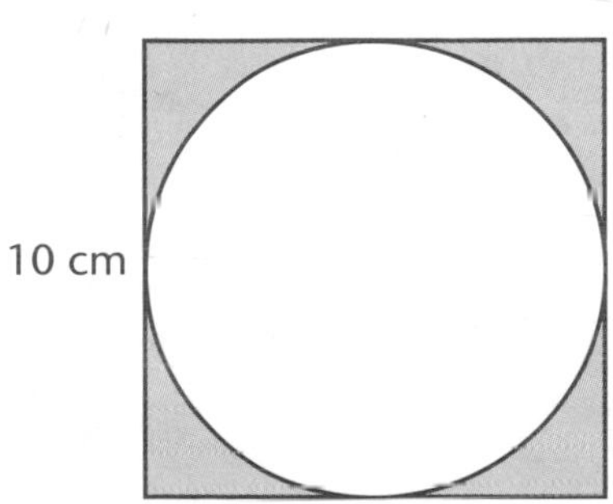

____________________ cm² **1 mark**

Volume

levels 5-6

1 **a** What is the volume of this cuboid?

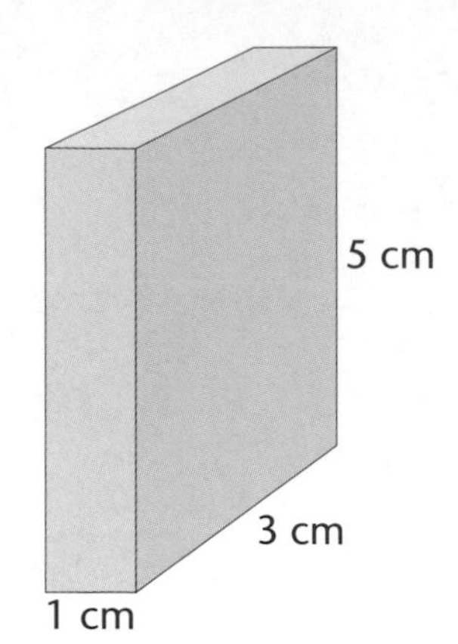

_______________ cm^3

1 mark

b What is the surface area?

_______________ cm^2

1 mark

2 A cuboid has a volume of 36 cm^3.
Its length is 6 cm and its width is 3 cm.
What is the height of the cuboid?

_______________ cm

3 A cuboid has a volume of 200 cm^3.
Its length and width are 5 cm.
What is the surface area?
Remember to include the units in your answer.

____

4 The volume of a cube is 64 cm^3.
What is the length of each edge of the cube?

_______________ cm

5 The surface area of this cuboid is 184 cm^2.
Work out the length of the cuboid.

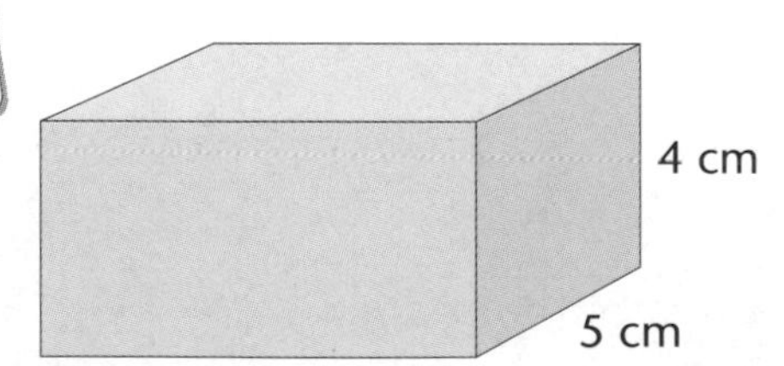

_______________ cm

1 mark

6 This is a net of a cuboid.
What is the volume of the cuboid?

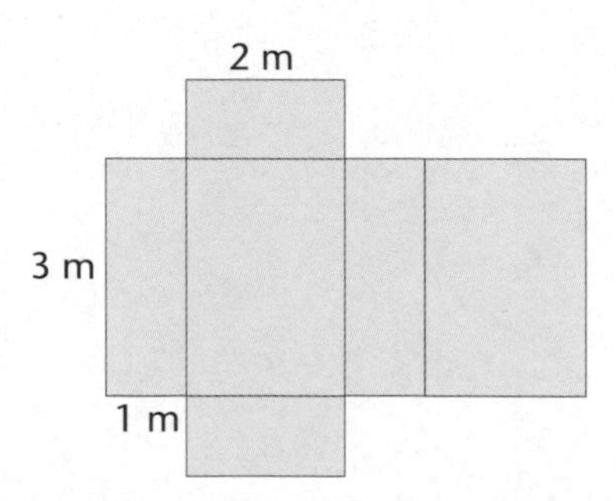

_______________ m^3

1 mark

7 Here are four cuboids.

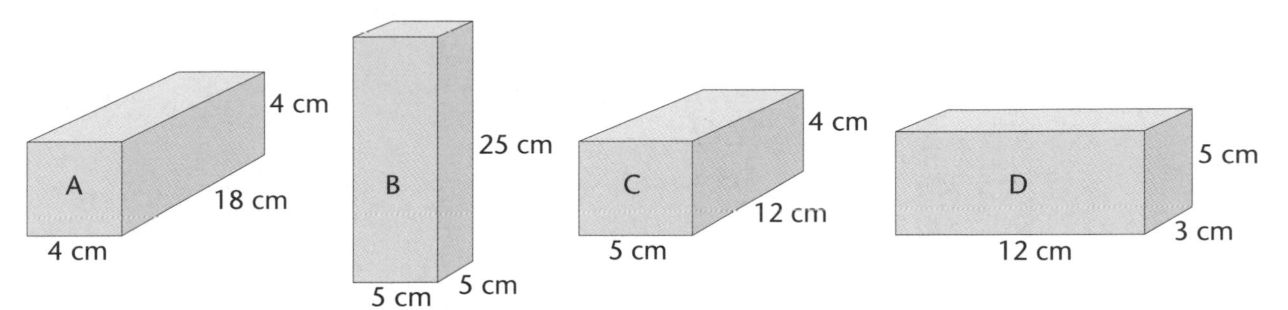

Rearrange the cuboids in the order of their volume, with the smallest first.

____

8 A tank has the following measurements. How many litres of water can it hold?

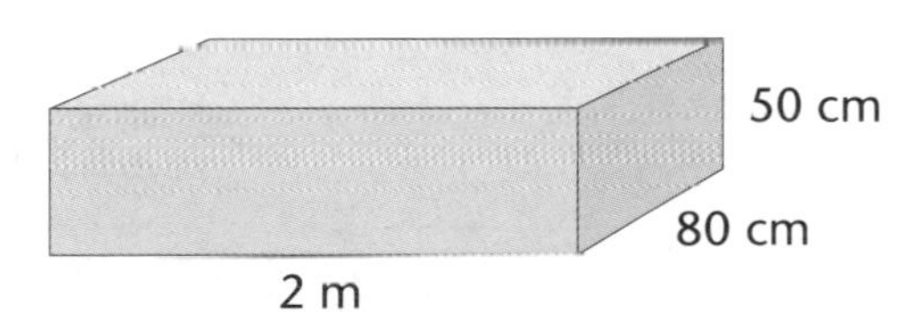

________________ litres

9 These two cuboids have the same volume. What is the value of x?

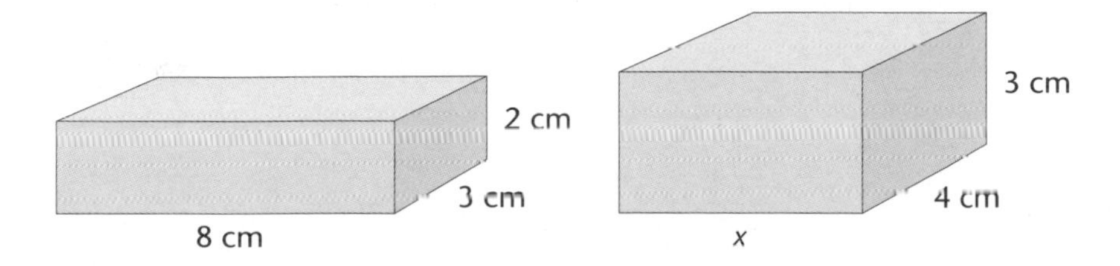

________________ cm

Constructions

1 Construct an angle of 60° at the point A on the line AB.

2 Construct the perpendicular bisector of the line AB.

3 Construct the angle bisector of the angle ABC.

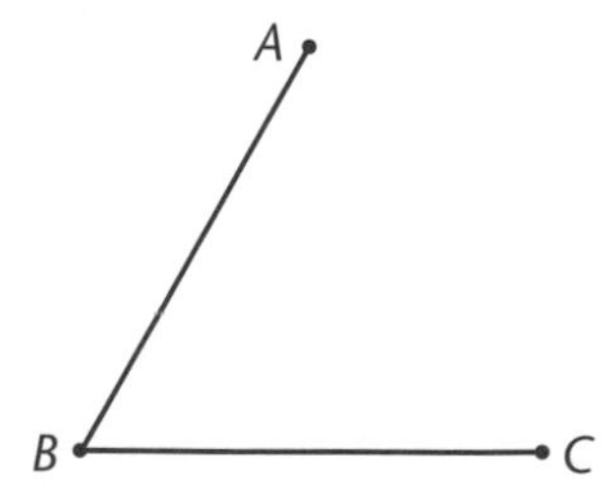

4 Draw this triangle accurately.

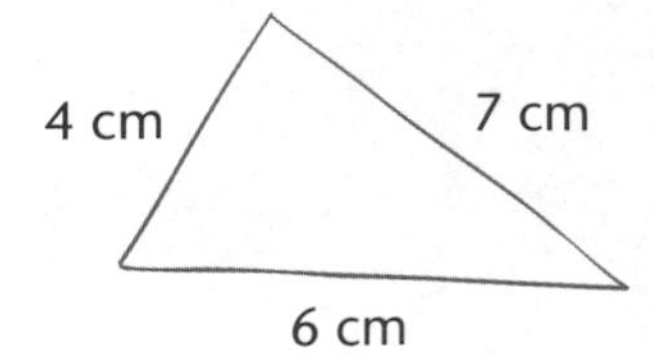

levels
6-7

5 Draw this triangle accurately.

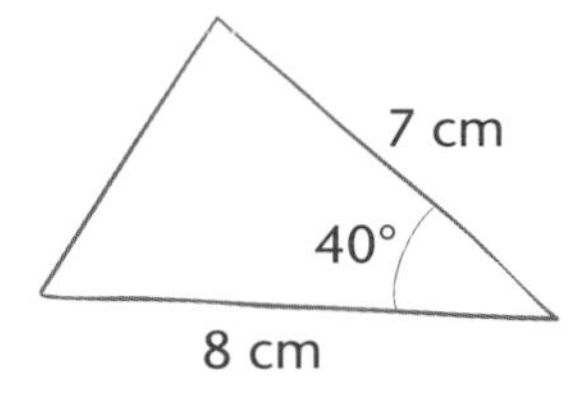

2 marks

6 Draw this triangle accurately.

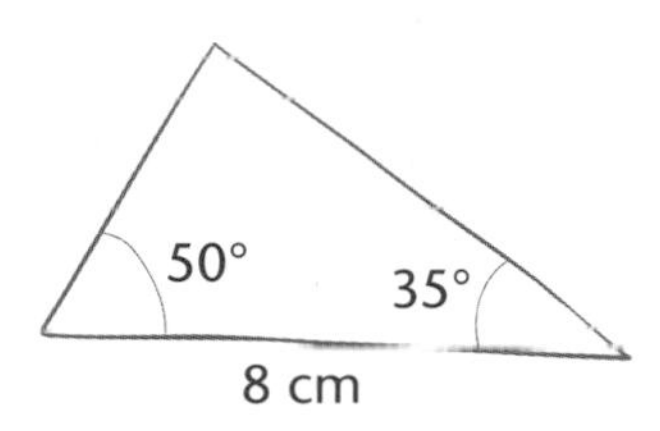

2 marks

Loci

levels 7-8

1 In each of these squares shade the region described.

a

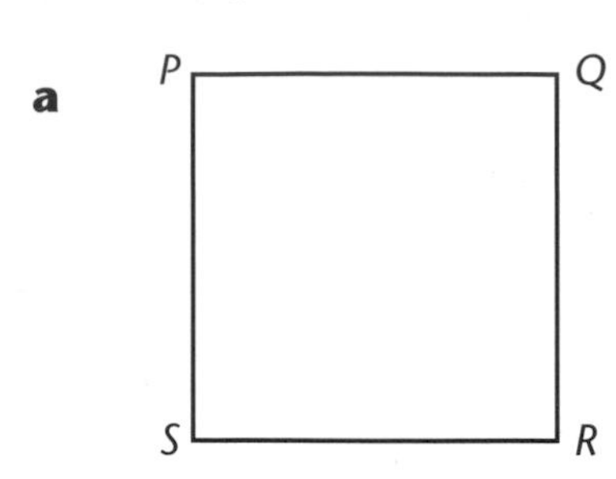

b

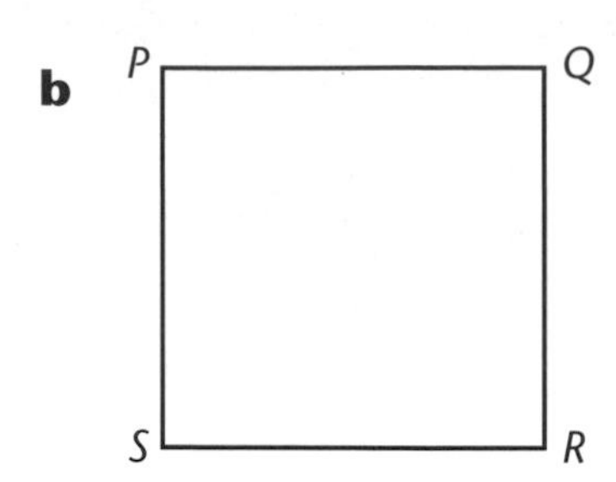

a All points that are nearer to *P* than to *Q*.

b All points that are nearer to *S* than to *Q*.

2 *ABCD* are squares of side 3 cm. Match the given loci to the diagrams.

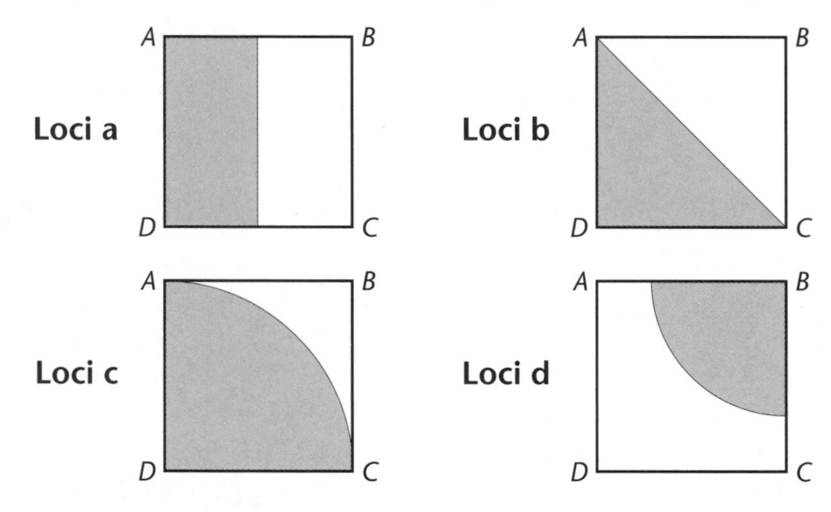

i All points nearer to *D* than to *B*. ____________________

ii All points within 3 cm of *D*. ____________________

iii All points nearer to the line *AD* than the line *BC*. ____________________

iv All points within 2 cm of *B*. ____________________

3 Construct the locus of the point that is the same distance from the lines *AB* and *AC*.

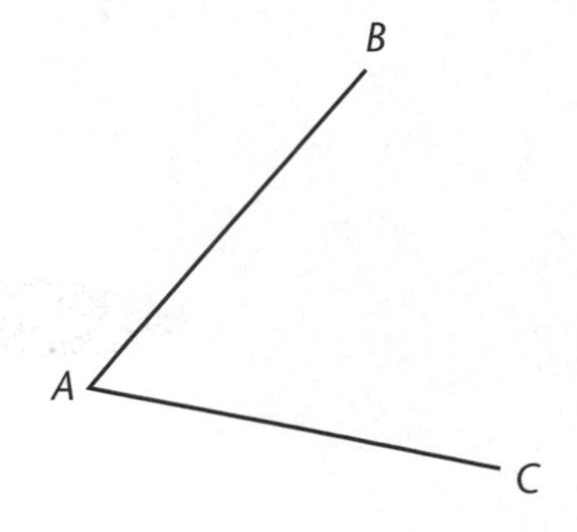

2 marks

levels 7-8

4 The diagram shows an island with two airports A and B.
The scale is 1 cm represents 10 km.
A radar station at A picks up aircraft within 30 km.
A radar station at B picks up aircraft within 40 km.

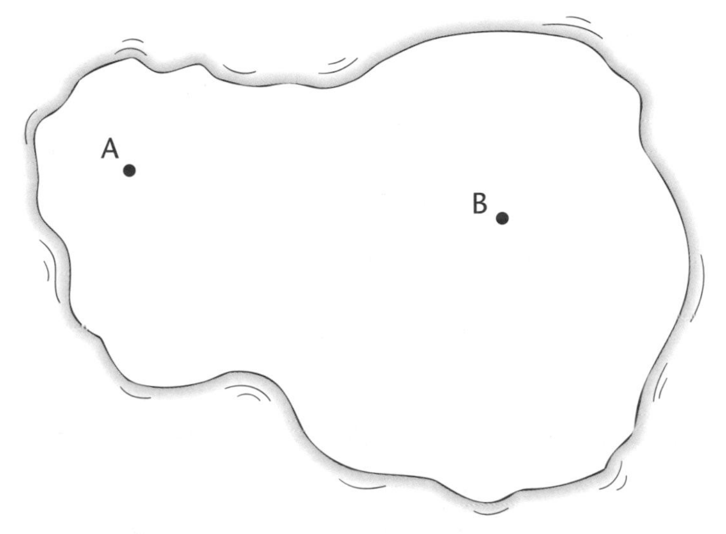

a Does the radar station at B pick up an aircraft flying directly over A? ______________

b Show all the points where aircraft are picked up by both radar stations.

5 The diagram shows a garden with a garden shed.
Each grid square represents 50 cm.

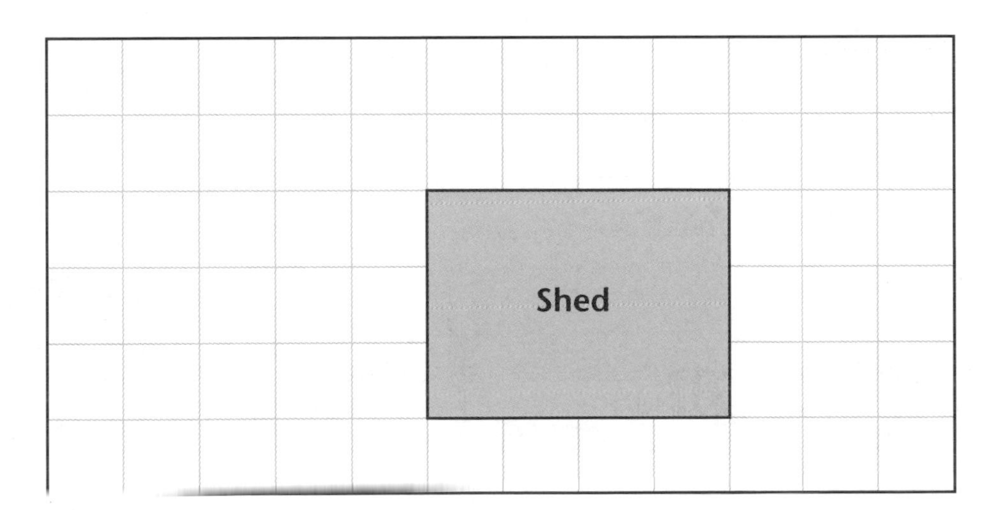

A tree is to be planted. It must not be planted within 1 m of the edge of the garden or the shed.
Shade clearly the area in which the tree can be planted.

1 mark

Similarity

levels 7-8

1 Look at the five triangles below. All lengths are centimetres.

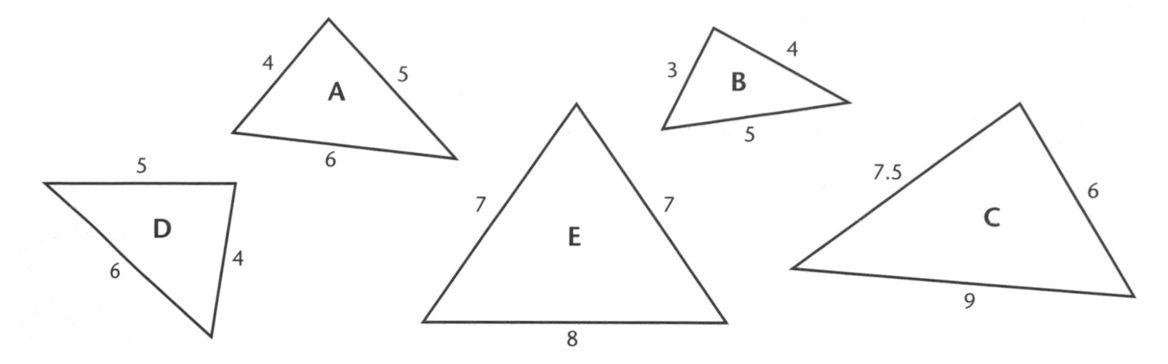

a Which two triangles are congruent to each other? Explain why.

____ 1 mark

b Which two triangles are similar to each other but not congruent? Explain why.

____ 1 mark

2 These two triangles are similar.

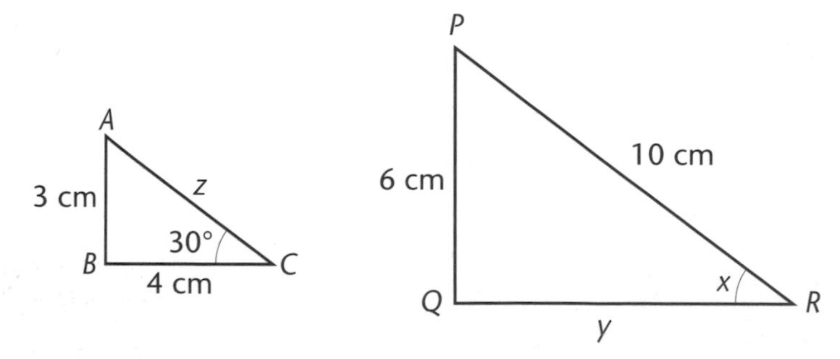

a What is the scale factor between triangle *ABC* and triangle *PQR*? ____________ 1 mark

b Write down the value of angle x. ____________ ° 1 mark

c Write down the length of the side y. ____________ cm 1 mark

d Write down the length of the side z. ____________ cm 1 mark

levels 7-8

3 Look at the two similar quadrilaterals, *ABCD* and *PQRS*.

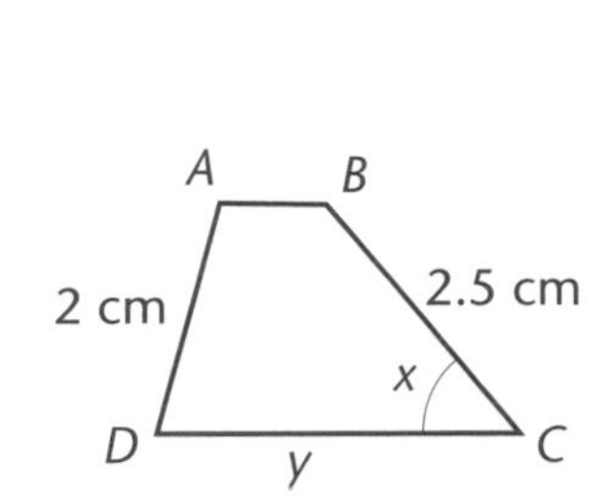

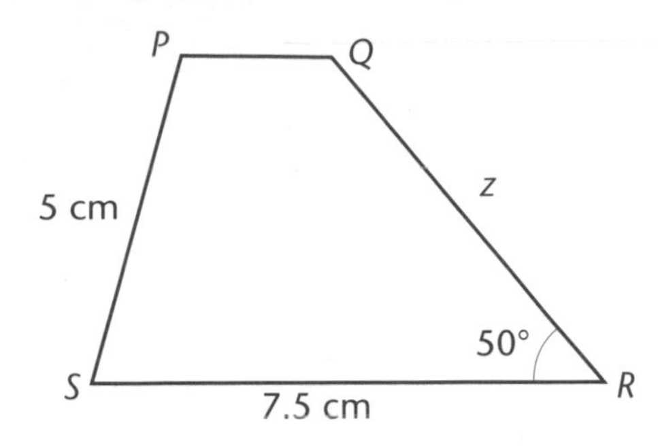

a What is the scale factor between quadrilaterals *ABCD* and *PQRS*? ________________ **1 mark**

b Write down the value of angle x. ________________ ° **1 mark**

c Write down the length of the side y. ________________ cm **1 mark**

d Write down the length of the side z. ________________ cm **1 mark**

4 Lines *AB* and *PQ* are parallel.

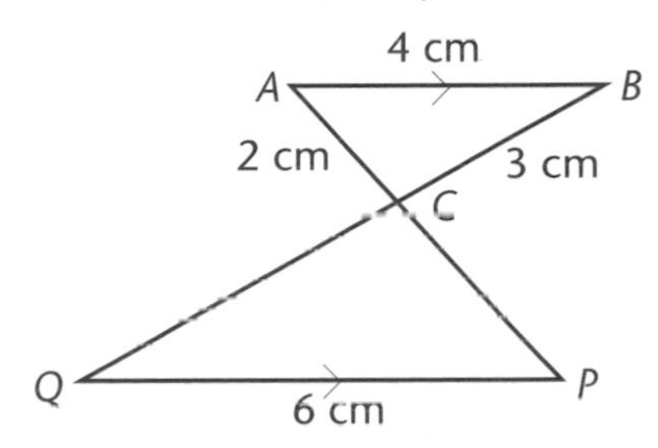

a Explain why triangles *ABC* and *CPQ* are similar. ________________

____

b Write down the value of the length *CP*. ________________ cm

c Write down the value of the length *CQ*. ________________ cm **3 marks**

5 Below are three tins of different sizes.

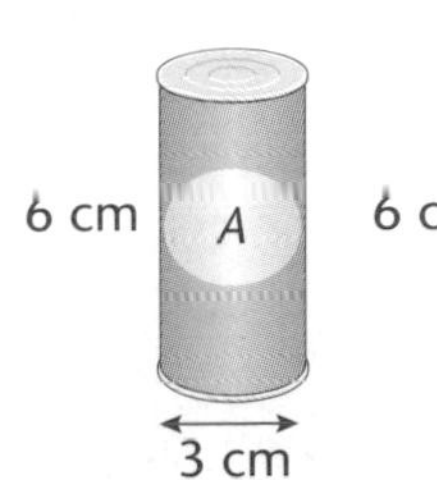

6 cm B 4 cm

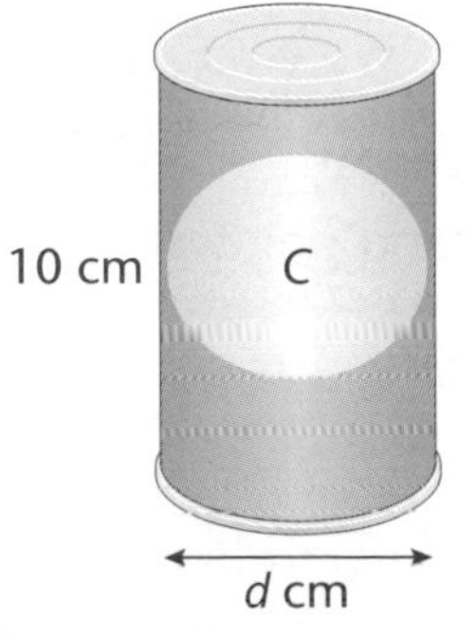

a Explain why tin *A* is not similar to tin *B*. ________________ **1 mark**

b Tin C is similar to tin *B*. Find the diameter, d, of tin C. ________________ cm **1 mark**

Pythagoras' theorem

1 A square of side length 5 cm has been drawn inside a circle.

a What is the area of the square?

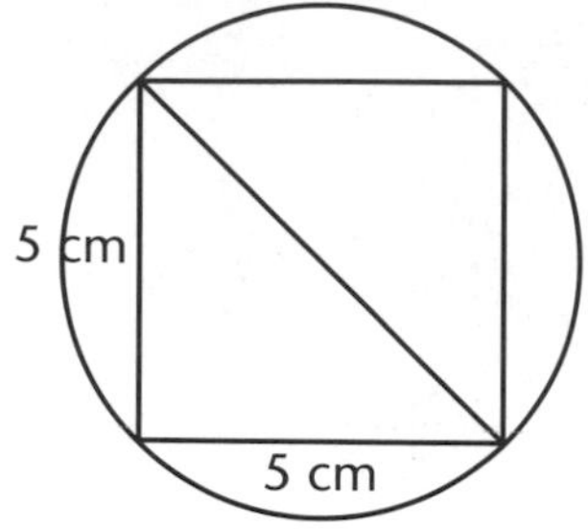

_______________ cm^2

1 mark

b Work out the diameter of the circle.

_______________ cm

c Work out the area of the circle.
Give your answer to 3 significant figures. _______________ cm^2

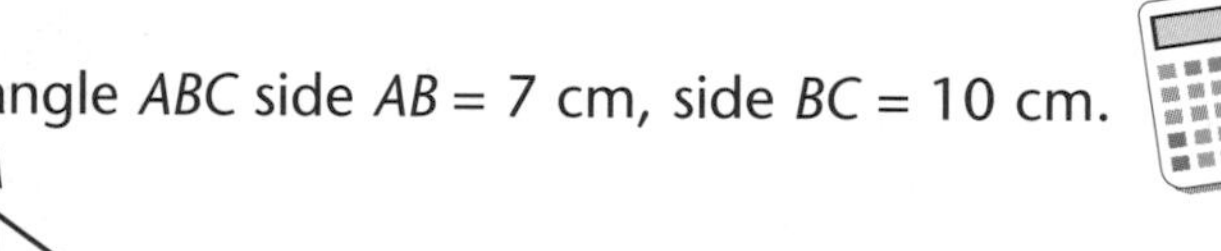

2 In triangle *ABC* side *AB* = 7 cm, side *BC* = 10 cm.

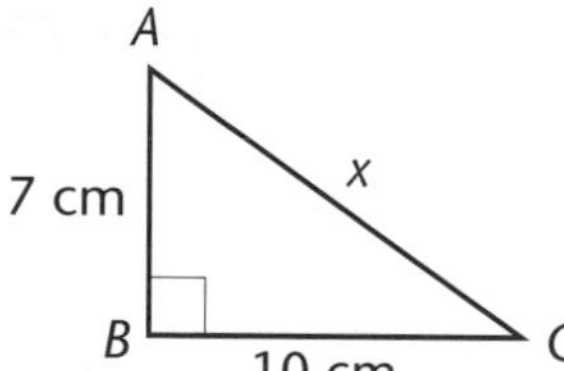

Calculate the length of the side *AC*.

Give your answer to 3 significant figures. _______________ cm

3 In triangle *ABC* side *AB* = 6 cm, side *AC* = 12 cm.

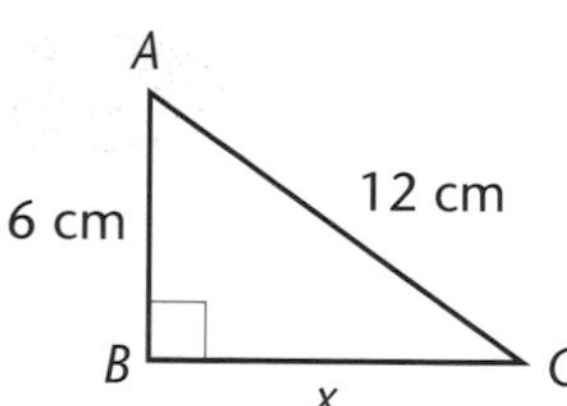

Calculate the length of the side *BC*.

Give your answer to 3 significant figures. _______________ cm

2 marks

4 A plane flies 30 km east and 20 km north.
It then flies directly back to its starting point.

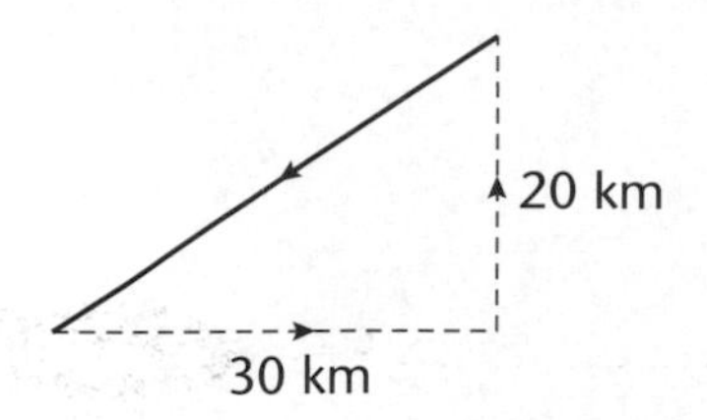

Calculate the total distance travelled. _______________ km

2 marks

level 7

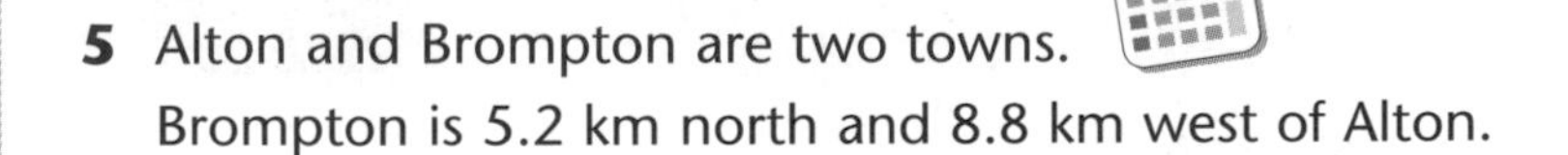

5 Alton and Brompton are two towns.
Brompton is 5.2 km north and 8.8 km west of Alton.

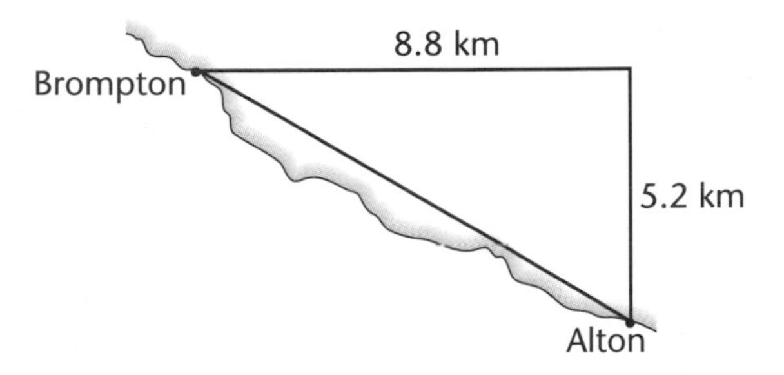

Calculate the direct distance from Alton to Brompton.

_______________ km

6 Triangle *ABC* has sides of 1.8 cm, 8 cm and 8.2 cm.

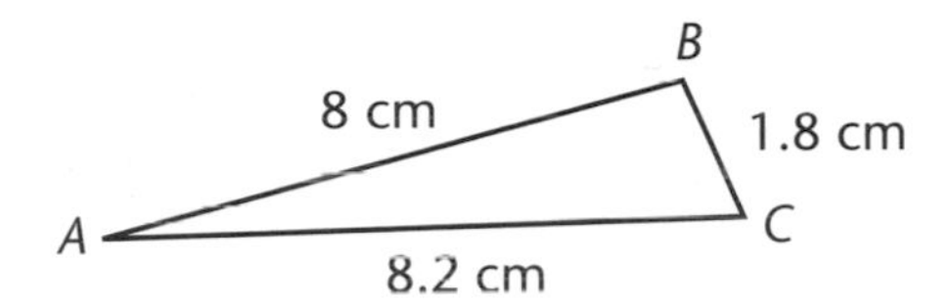

Is triangle *ABC* right-angled?
Justify your answer. _______________________________________

7 A circle with a diameter of 10 cm is drawn within a square.

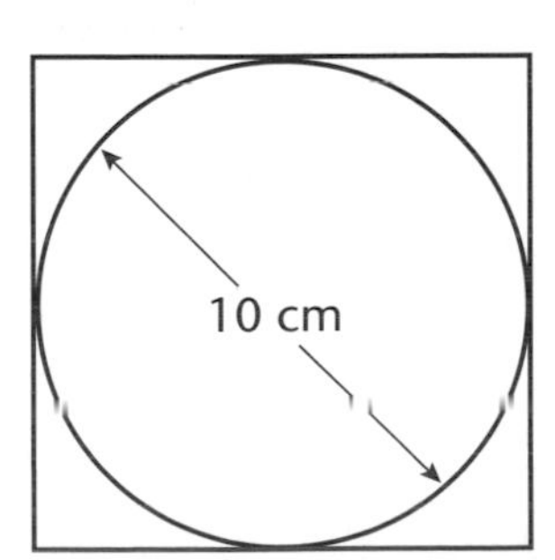

Calculate the length of the square's diagonal.
Give your answer to 3 significant figures.

_______________ cm

SHAPE, SPACE AND MEASURES

Trigonometry

level 8

1 Calculate the length x in this right-angled triangle.

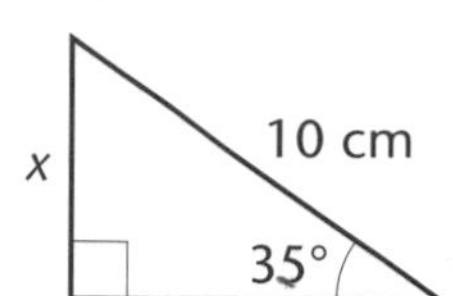

__________ cm 

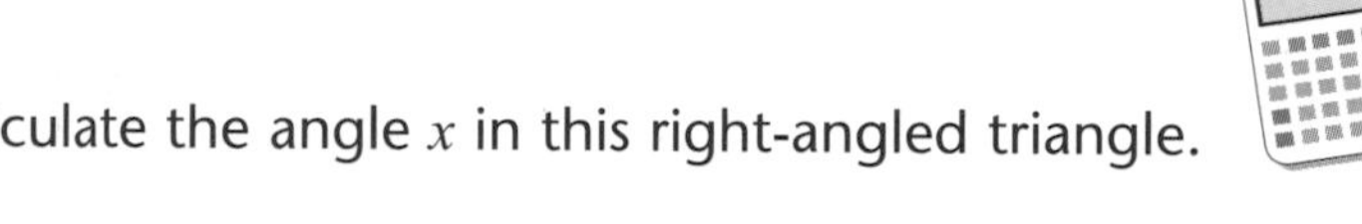

2 Calculate the angle x in this right-angled triangle.

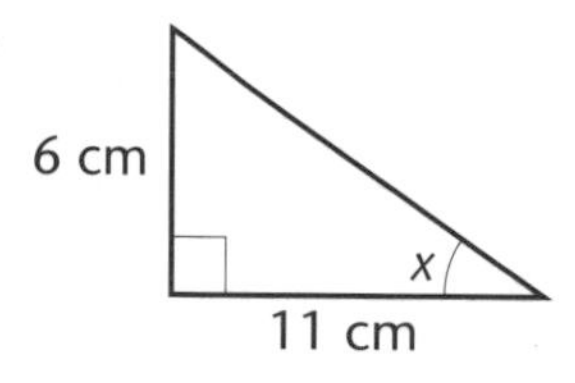

__________ degrees 

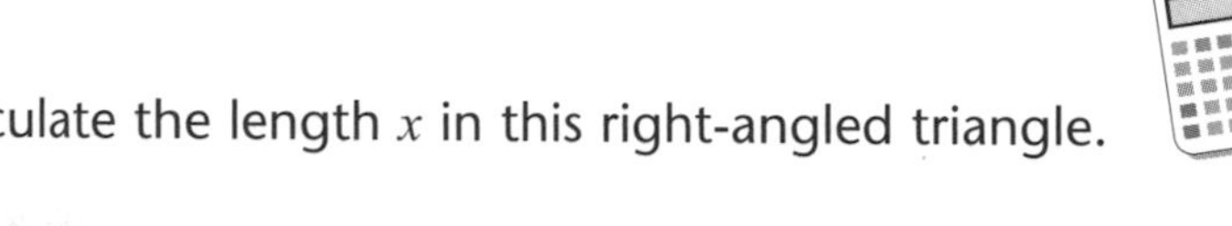

3 Calculate the length x in this right-angled triangle.

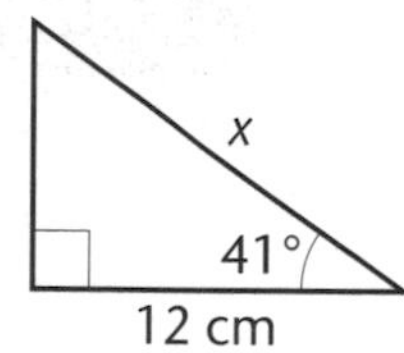

__________ cm 

4 In triangle ACB side $AB = 6$ cm, side $AC = 12$ cm.

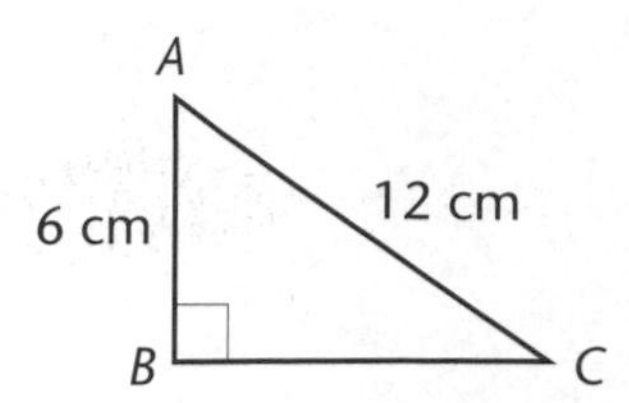

Calculate the size of the angle ACB.

__________ degrees 

level 8

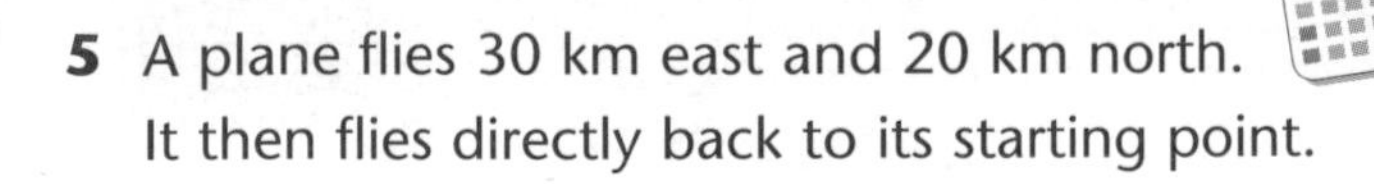

5 A plane flies 30 km east and 20 km north.
It then flies directly back to its starting point.

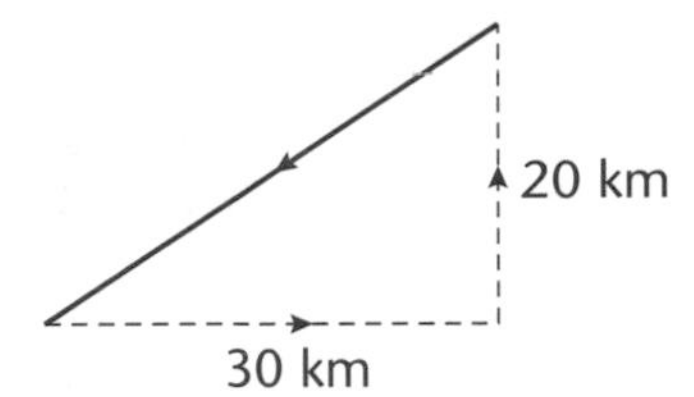

Calculate the bearing of the return flight. ____________________ degrees

6 Alton and Brompton are two towns.
Brompton is 5.2 km north and 8.8 km west of Alton.

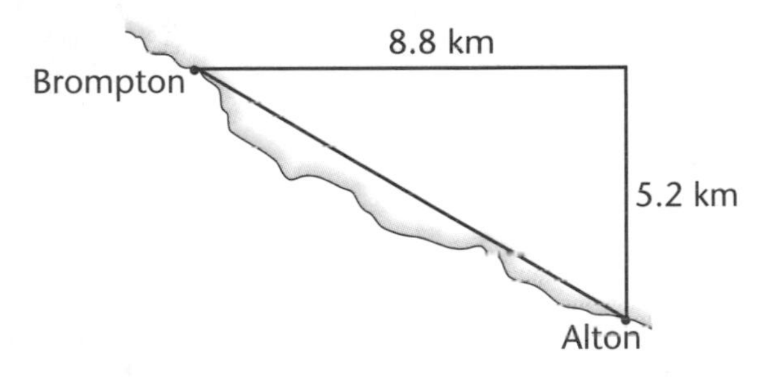

Calculate the bearing of Alton from Brompton. ____________________ degrees

2 marks

7 *ABC* is an isosceles triangle with sides of 6 cm, 8 cm and 8 cm.

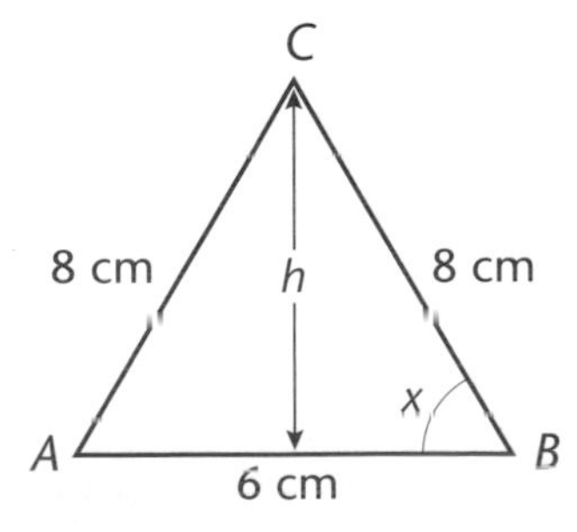

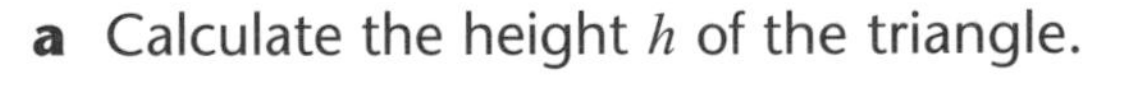

a Calculate the height h of the triangle.

____________________ cm

b Calculate the angle marked x.

____________________ degrees

c Calculate the area of the triangle.

____________________ cm^2

Sectors and circle theorems

level 8

1 A circle has the same numerical value for the circumference as the area. 
What is the radius?

_______________ cm 

2 A sector of a circle radius 6 cm has an angle of 72°. 

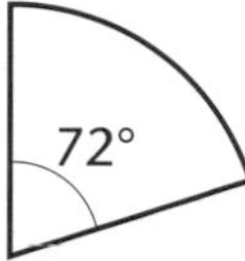

a What is the perimeter of the sector?

 cm 

b What is the area of the sector?

_______________ cm² 

3 **a** What fraction of 360° is **i** 90°_______________ **ii** 80°_______________  

b The diagram shows two sectors of circles A and B.

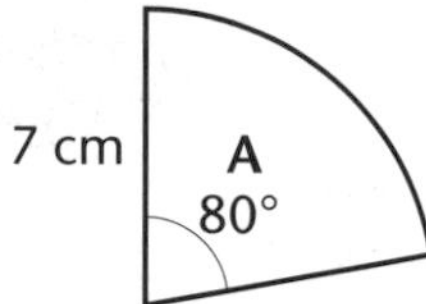

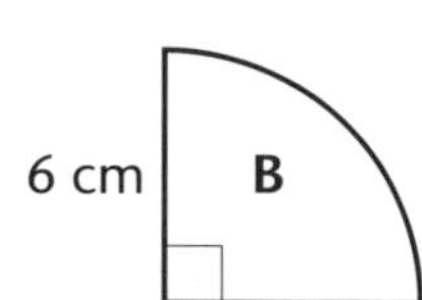

Which sector has the bigger area? _______________ 

c Which sector has the bigger perimeter?

4 Cylinder A has a diameter of 15 cm and a height of 20 cm.
Cylinder B has a diameter of 20 cm and a height of 15 cm. 

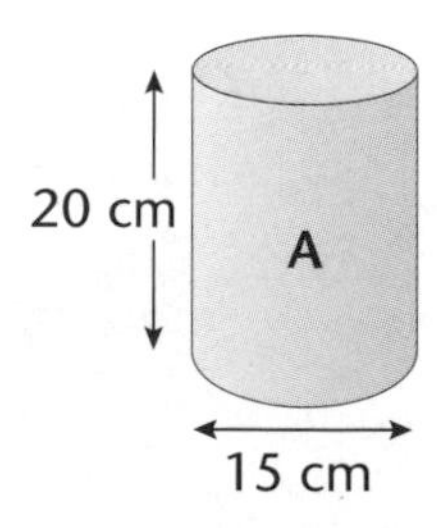

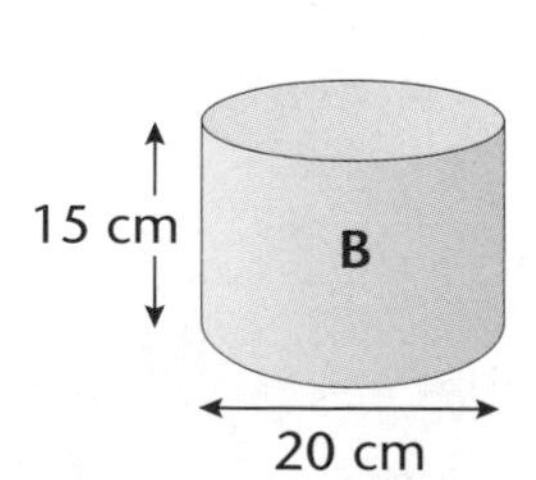

a Which cylinder has the greater surface area?

_______________ 

b Which cylinder has the greater volume?

_______________ 2 marks

level 8

5 Write down the value of the angles marked with letters in these diagrams.

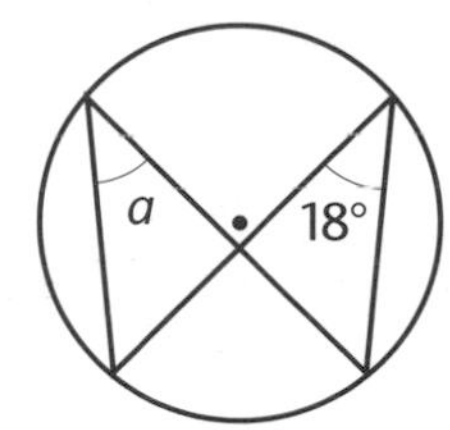

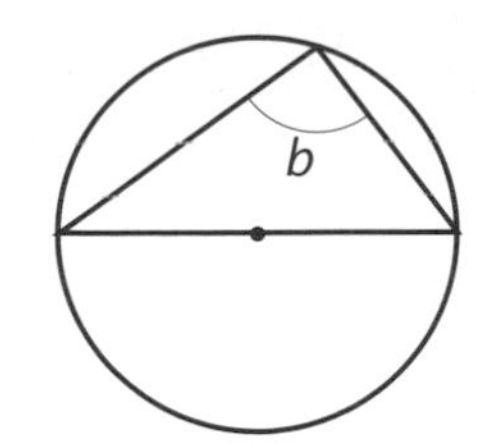

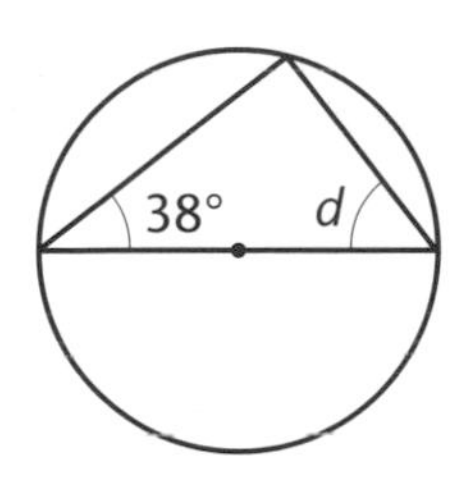

a = ____________ degrees

b = ____________ degrees

c = ____________ degrees

d = ____________ degrees

6 Write down the value of the angles marked with letters in these diagrams.

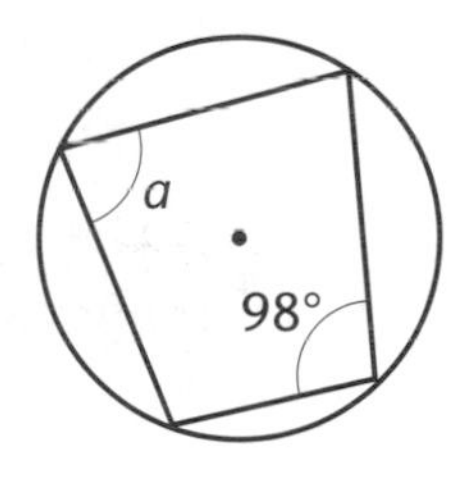

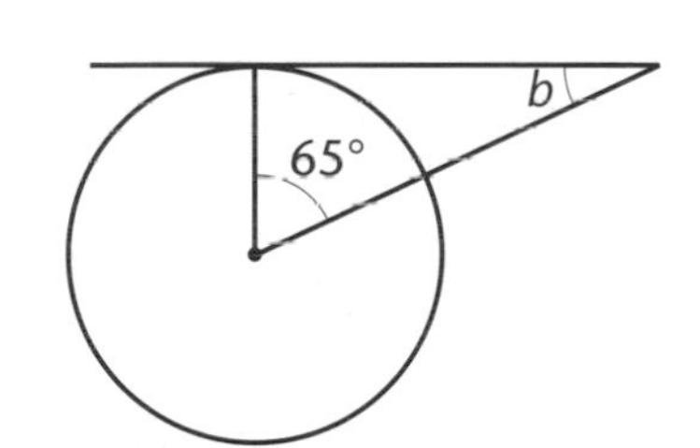

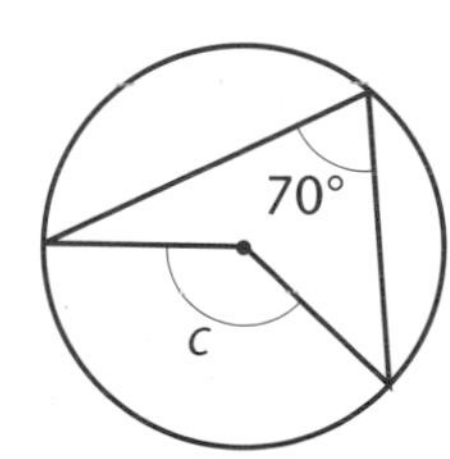

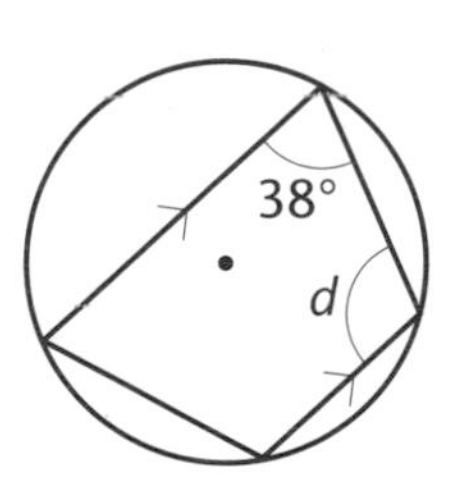

a = ____________ degrees

b = ____________ degrees

c = ____________ degrees

d = ____________ degrees

HANDLING DATA Comparing distributions

levels 5-6

1 This grouped frequency table shows the ages of 50 members of a tennis club.
Which of the following statements could be true or must be false?

Age	Frequency
21 – 30	15
31 – 40	18
41 – 50	12
51 – 60	4
61 – 70	1
Total	50

	Could be true	Must be false
a The range of the ages is 49	☐	☐
b The median age is 45	☐	☐
c The modal age is 65	☐	☐
d The modal age is 39	☐	☐

4 marks

2 The bar charts show the number of days absent in a week for students in two different classes in Year 10.

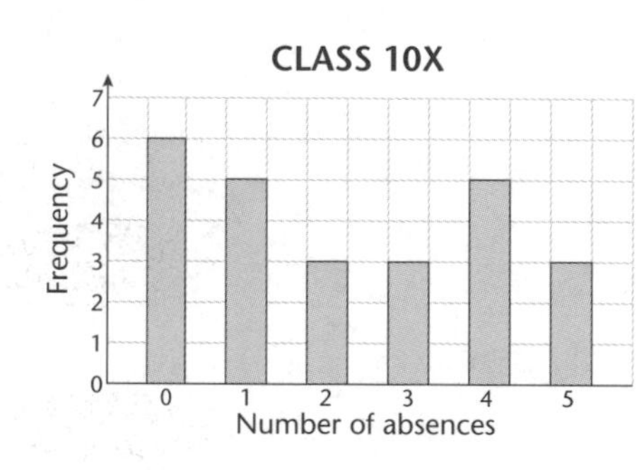

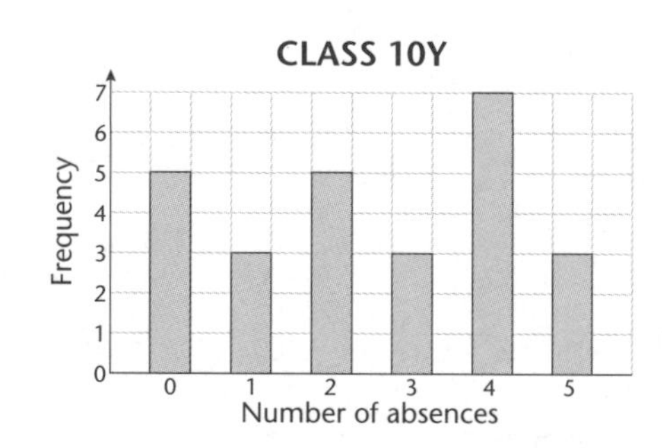

a State the modal number of absences for each class. 10X ______ 10Y ______ **2 marks**

b Work out the median number of absences for each class. 10X______ 10Y______ **2 marks**

c Work out the mean number of absences for each class. 10X ______ 10Y ______ **2 marks**

d Which class was worse for absentees? Give a reason for your answer.

____

____ **1 mark**

3 The table shows information about how late two buses are over a 20 day period.
The data is in minutes.

	Mean	Median	Mode	Range
Bus A	8	9	3	15
Bus B	8	4	0	10

Which bus is more reliable?
Give a reason for your answer.

____

____ **2 marks**

levels 5-6

4 The following data shows the weekly wages in a small factory with eight workers.

£95 £220 £220 £220 £220 £220 £320 £700

Everyone gets a £20 a week pay rise.

Which of the following is true for the new wages?

Tick the correct answer. (There may be more than one.)

a ☐ The mean will increase by £20.

b ☐ The mode will increase by £20.

c ☐ The median will increase by £20.

d ☐ The range will increase by £20. **1 mark**

5 Two girls want to be in the school senior netball team.

The number of goals they scored in their last 10 junior matches were

Aisha	3	7	2	4	4	1	1	0	2	1
Betty	3	4	2	3	3	1	2	2	3	2

a Work out the mean number of goals for each girl. ____________ **2 marks**

b Work out the range for each girl. ____________ **2 marks**

c Which girl should be chosen for the senior team and why?

____________ **1 mark**

6 The data shows the number of tomatoes from 10 plants grown in a greenhouse and 10 plants grown outside.

Greenhouse	2	8	7	12	4	6	9	10	8	4
Outside	5	5	8	9	8	6	6	9	8	6

a Work out the mean number of tomatoes per plant for

i the greenhouse ____________

ii outside ____________ **2 marks**

b Work out the range for the number of tomatoes per plant for

i the greenhouse ____________

ii outside ____________ **2 marks**

c Which is the better place to grow tomatoes and why?

____________ **1 mark**

HANDLING DATA Line graphs

levels 5-6

1 The graph shows the trend in the temperature in a garden over a week in May. The readings were taken at midday each day.

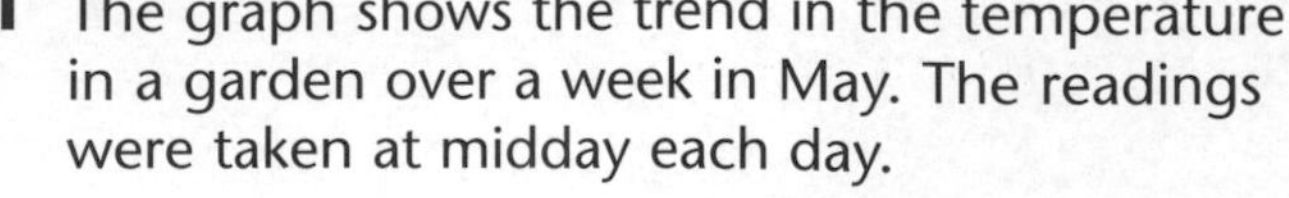

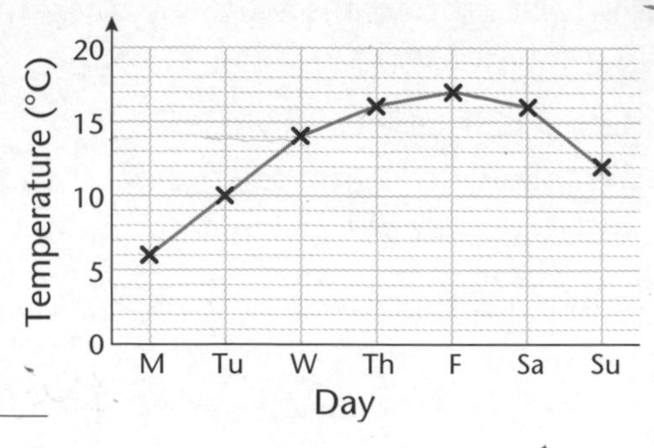

a What was the temperature on Wednesday?

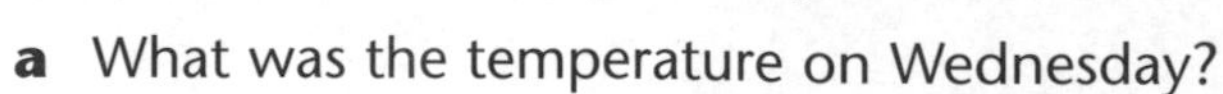

__________________ **1 mark**

b The temperature was 16 °C on 2 days.

Which days? ____________ and ____________ **1 mark**

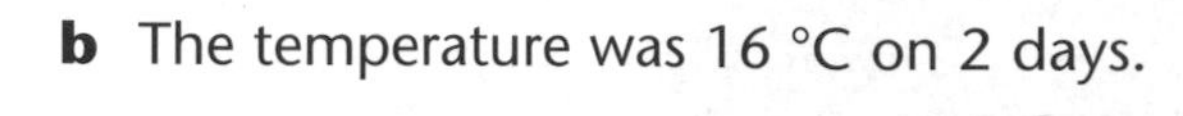

c On what day was the temperature highest? ____________ **1 mark**

d Explain why you cannot tell what the temperature was at midnight on Wednesday. ________________________

____ **1 mark**

2 The graph shows the trend in the temperature in a garden over a week. The temperatures were recorded at 12 midday and 12 midnight.

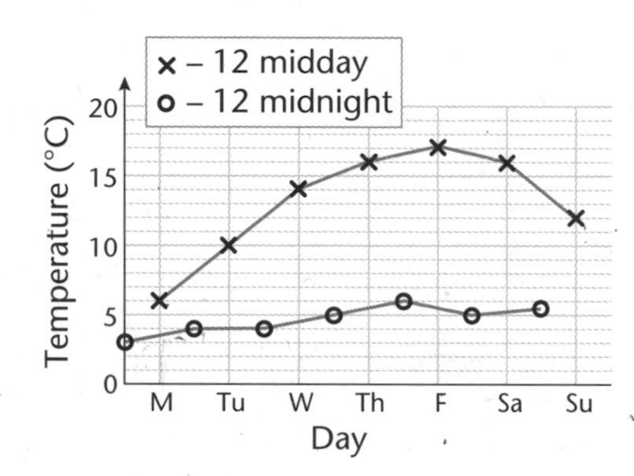

a What was the temperature at midnight and midday on Monday?

midnight __________ midday __________ **1 mark**

b What was the difference in temperatures between midday and midnight on Tuesday? ____________________ **1 mark**

c Which day had the greatest difference between the temperatures at midday and midnight? ____________ **1 mark**

3 The graph shows the miles travelled each month by a lorry driver.

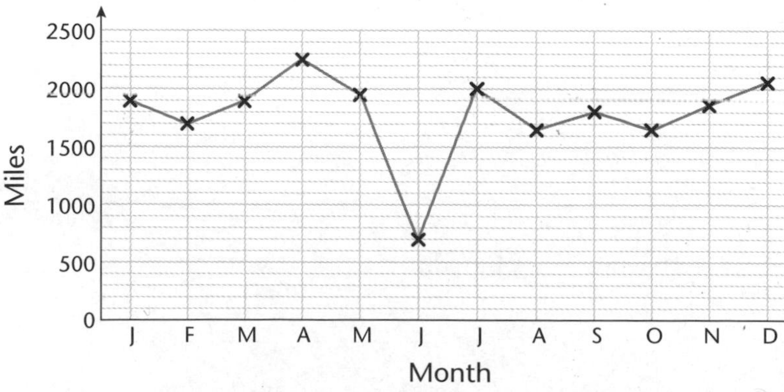

a What was the distance he travelled in January? ____________ **1 mark**

b What was the total distance travelled in the first three months of the year?

__________________ **1 mark**

c Which month was he most likely to be on holiday? ________________ **1 mark**

levels 5-6

4 The graph shows the results of an experiment to see if a detergent has any effect on bacteria. Results were recorded every hour from 10 am to 4 pm.

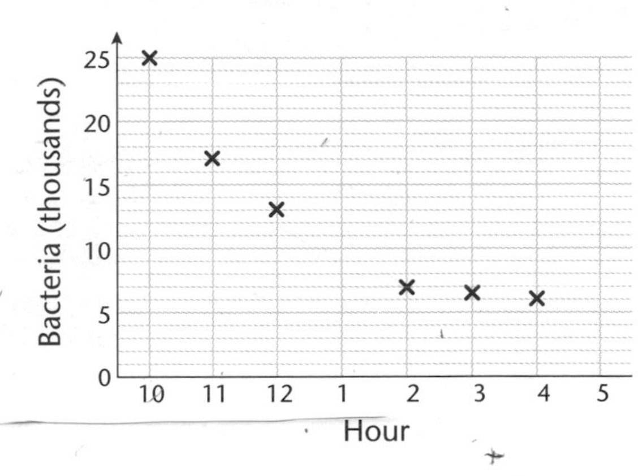

a The scientist was at lunch at 1 pm.
Estimate the number of bacteria at 1 pm.
Give your answer to the nearest 100. ____________

1 mark

b When the number of bacteria drop below 5000 the detergent is said to be effective.
The scientist claimed that this graph shows that this detergent is effective.
Is this claim true or false? Give a reason for your answer.

____

1 mark

c Is it possible to estimate the number of bacteria at 11.30 am?
Give a reason for your answer.

____

1 mark

5 The graph shows the depth of water in a drain during a heavy rainstorm.
When the depth gets to 30 cm the area will flood.
John predicts that the area will flood by 2 pm.
Is this claim justified? Explain your answer.

____

____

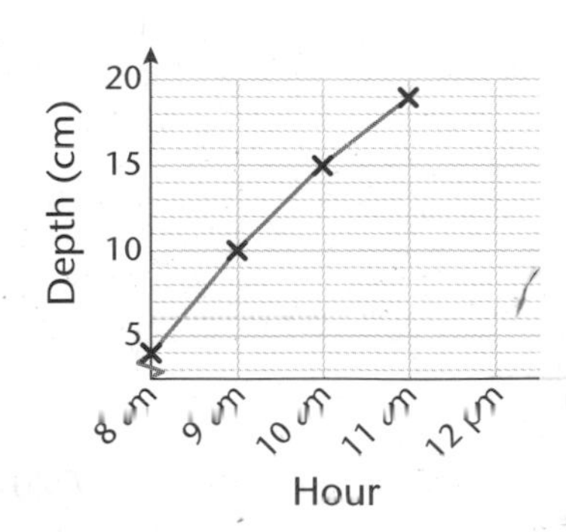

1 mark

Statistical diagrams

level 6

1 Look at the frequency diagram which compares the heights of men and women in a church choir.
Comment on the differences in the distributions between the men and the women.

____

____

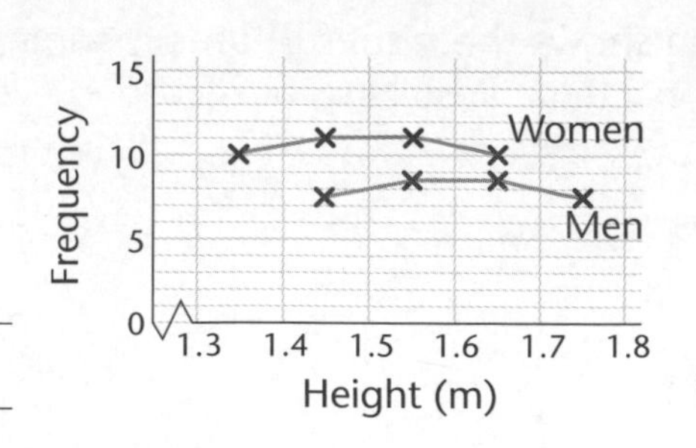

2 marks

2 The data shows the number of e-mails received over 15 days.

7, 12, 22, 17, 11, 9, 8, 13, 15, 21, 19, 18, 8, 8, 13

Show the data in a stem-and-leaf diagram using the key 1 | 2 represents 12.

1 mark

3 The stem-and-leaf diagram shows the ages of 12 members of a chess club.

Stem	Leaves						
1	3	8	9				
2	0	2	2	2	3	5	8
3	1	4					

Key: 1 | 3 represents 13 years.

a How old is the oldest member? ________________

1 mark

b What is the modal age of the members? ________________

1 mark

c What is the range of the ages of the members? ________________

1 mark

4 The table shows information on colours of cars in a car park.

Colour	Frequency	Angle
Blue	9	
White	4	
Silver	7	
Total	20	

a Complete the column for the angle that each colour would have on a pie chart.

b Draw a pie chart to show the information.

2 marks

5 Look at the pie chart which shows the favourite drinks of some people.
48 students chose coffee.
How many students altogether were in the survey?

____

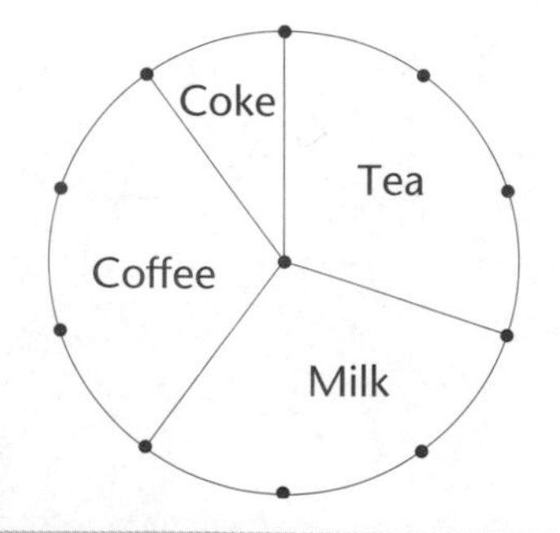

6 The pie chart shows the results of an election survey.
It is not drawn accurately.
120 people said they would vote Labour.
How many people said they would vote Green?

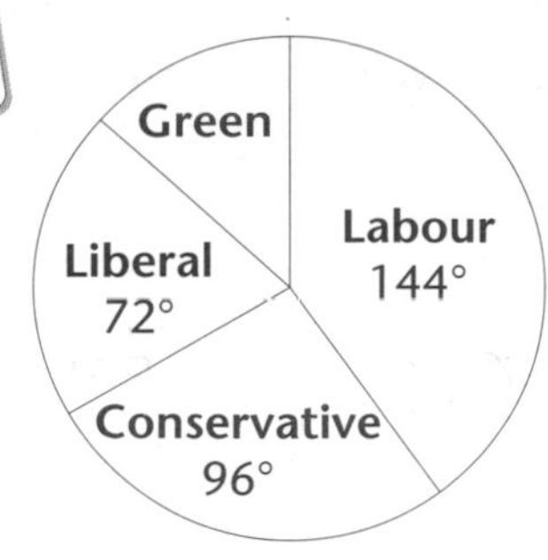

____

1 mark

7 Which of the following could not be the angles of the sectors in a pie chart?

Tick the correct answer.

a ☐ 90°, 60°, 130°, 80°   **b** ☐ 75°, 25°, 200°, 60°

c ☐ 62°, 144°, 96°, 48°   **d** ☐ 90°, 90°, 90°, 90°

1 mark

8 Phyllis did a survey about the month people were born in.
She surveyed 240 people.
Which of the following would be a valid reason why a pie chart is not a good method of representing the data?
Tick the correct answer.

a ☐ There are too many sectors to show a valid comparison.

b ☐ 240 doesn't divide into 360 exactly.

c ☐ People might lie about their birthday.

d ☐ You couldn't fit the labels on the pie chart.

1 mark

9 The table shows information about the nationality of people on a plane.

	British	American	French	German
Percentage	45%	25%	20%	10%

Draw a pie chart to represent the data.

10 Draw a pie chart to represent this data.

Blue	Silver	Black
12	6	2

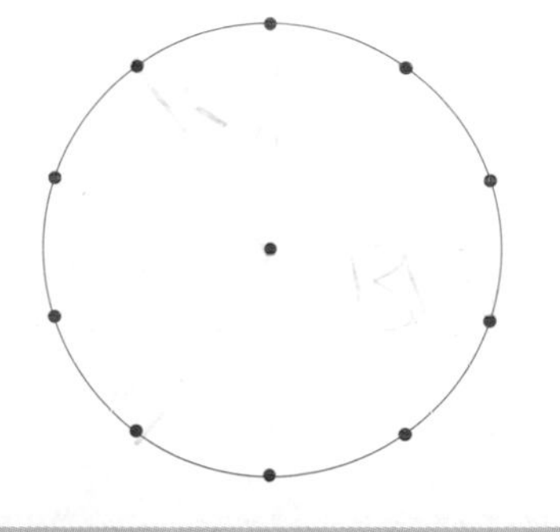

1 mark

Scatter diagrams

level 6

1 Correlation can be described using the following terms.

S Strong positive correlation | **W** Weak positive correlation

N No correlation

G Strong negative correlation | **K** Weak negative correlation

a Match each diagram with one of the descriptions above.

i

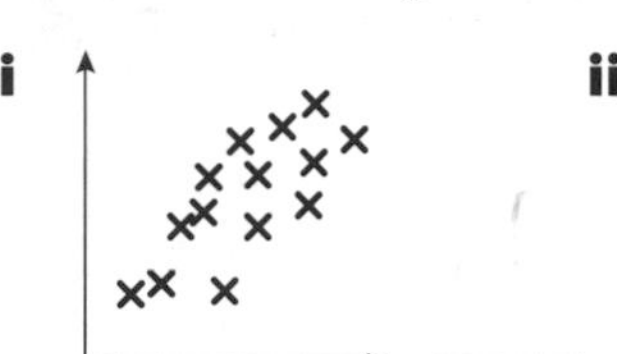

ii

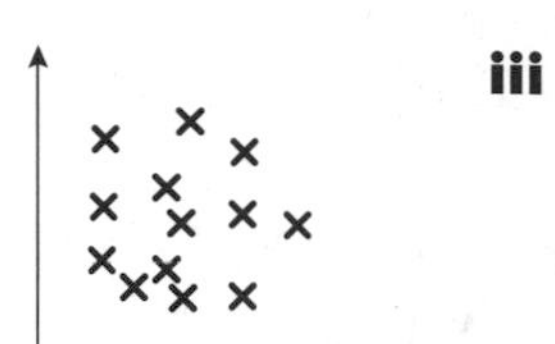

iii

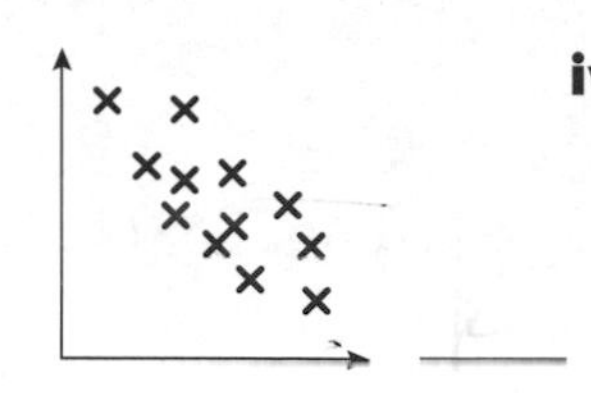

iv

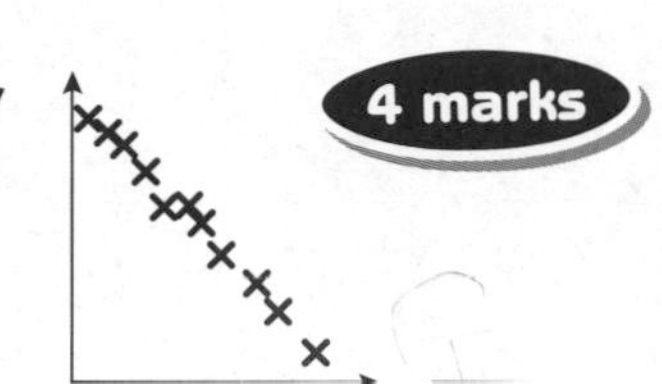

4 marks

b Match the types of correlation to these comparisons.

i The age of a car and its top speed. ____________

ii The number of men building a wall and the time taken to build it. ____________

iii The number of ice creams sold and the temperature. ____________

iv The value of cars and their age. ____________

4 marks

2 The scatter graph shows the heights and weights of a breed of horses.

a Draw a line of best fit on the data.

1 mark

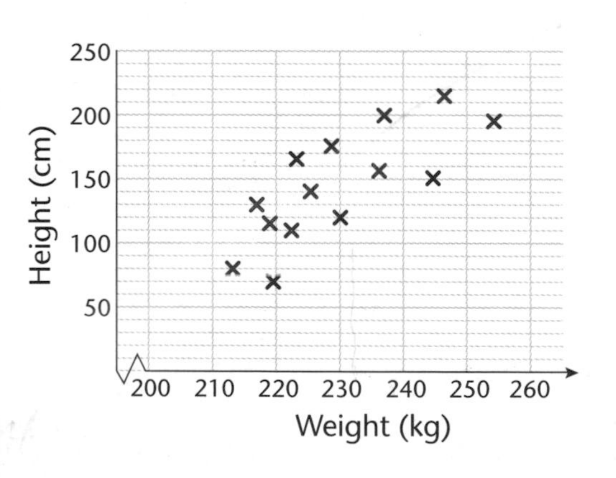

b A horse of the same breed has a weight of 232 kg. Estimate its height.

____

1 mark

c Another horse has a weight of 200 kg and is 200 cm tall.

Could this horse be of the same breed?

Give a reason for your answer.

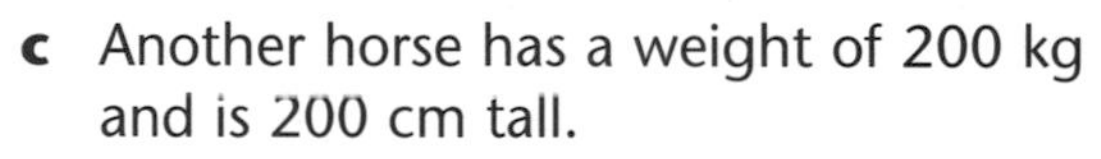

____

1 mark

3 The graph shows the finishing times of runners in a marathon and the number of miles run per week in training. A line of best fit has been drawn.

Neil runs 120 miles a week in training.

What is his likely finishing time?

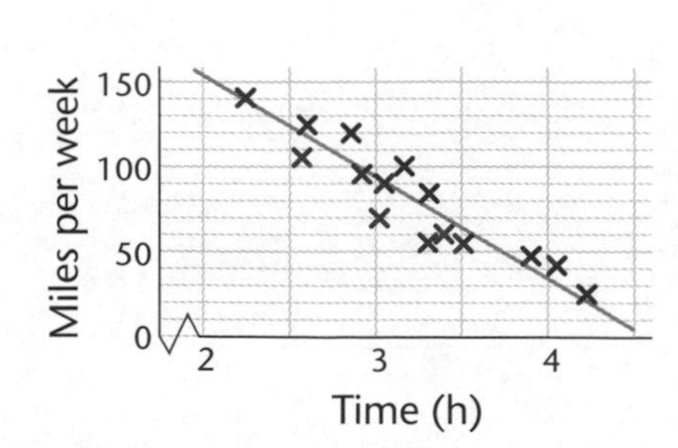

____

1 mark

level 6

4 The scatter graph shows the ages and number of years in the job for the men and women employed in a do-it-yourself store.

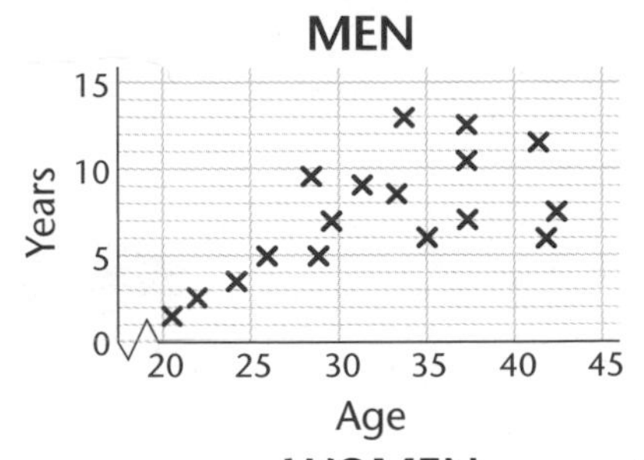

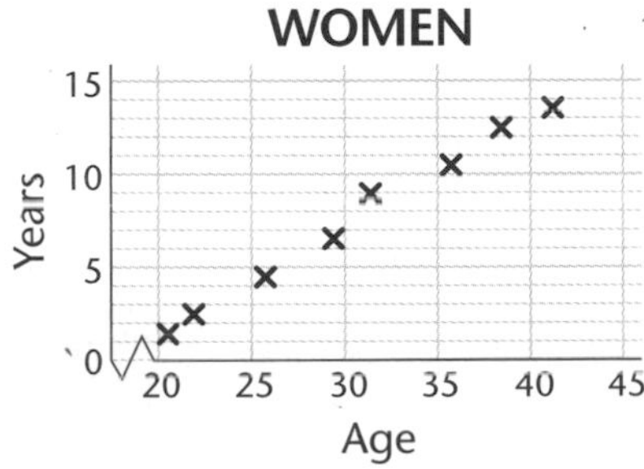

Which of the following statements is true for the data?
Tick the correct answer. (There may be more than one.)

a ☐ The women's scatter graph shows strong positive correlation.

b ☐ The men's scatter graph shows strong positive correlation.

c ☐ There are more men employed than women.

d ☐ For men over 30 there is no correlation between their age and the number of years employed.

1 mark

5 Three different variables are:

A the time it takes to plaster a wall

B the number of men working on a job

C the weekly wage bill

Which of the following will be true?
Tick the correct answer. (There may be more than one.)

a ☐ A and B will show negative correlation.

b ☐ A and C will show no correlation.

c ☐ B and C will show negative correlation.

d ☐ B and C will show positive correlation.

1 mark

6 The scatter graph shows the ages and finishing times in a marathon for 10 members of a running club.

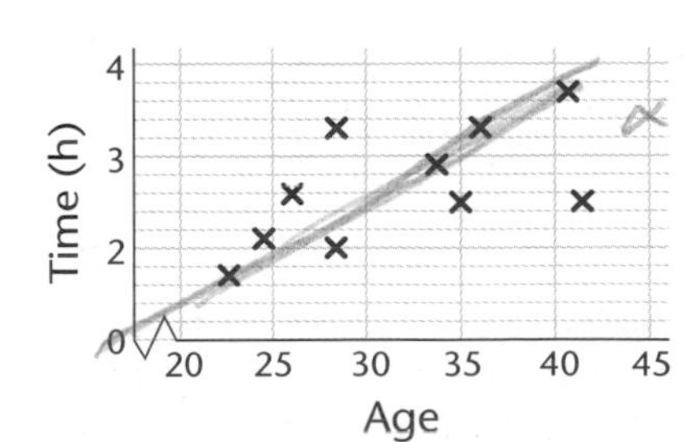

a Describe the correlation.

____

1 mark

b Draw a line of best fit on the data.

c Another member of the club is 45 years old.
What is his likely finishing time?

____

1 mark

Surveys

level 6

1 When carrying out a survey which of the following should you do?
Tick the correct answer. (There may be more than one.)

a ☐ Ask friends, relatives or neighbours.

b ☐ Ask a variety of people.

c ☐ Ask questions that are unbiased.

d ☐ Make sure there is an equal number of boys and girls.

1 mark

2 Jodie does a survey to find out people's views on the following question.
'Should school uniform be worn by all students?'

a Say why each of the following would not be good responses to this question.

i Yes ☐ No ☐

Criticism ______________________________

____

1 mark

ii Agree ☐ Don't know ☐ Disagree ☐

Criticism ______________________________

1 mark

iii Ties ☐ Blazers ☐ Shoes ☐ Caps ☐

Criticism ______________________________

1 mark

b Say why this is a good response to the same question.

Strongly agree ☐ Agree somewhat ☐ Neither agree nor disagree ☐ Disagree somewhat ☐ Strongly Disagree ☐

Reason ______________________________

1 mark

3 Give two reasons why this is not a good survey question.
'People who smoke are not very intelligent. Don't you agree?'

Reason 1 ______________________________

1 mark

Reason 2 ______________________________

1 mark

4 Four students are doing a survey on sport.
Asif decides to ask 30 students in the Badminton club.
Benny decides to ask his Year 9 tutor group.
Colin decides to ask 30 students on the school field at lunchtime.
Derek gets a list of all the students in school and randomly selects 30 names to ask.
Who will get the most reliable results? Give a reason for your answer.

Reason ______________________________

1 mark

level 6

5 What is wrong with this question on eye colour?

What is your eye colour?	**Brown** Yes/No	**Blue** Yes/No

Reason ______________________________ 1 mark

6 The headmaster gets an alphabetical list of all the students in the school and sends a questionnaire to every tenth name on the list.
Explain why this will give a good sample of the students.

Reason ______________________________ 1 mark

7 In a clothing factory there are 100 women employees and 15 men employees.
The managing director sends a questionnaire to the men and 15 of the women.
Explain why this will not give a representative sample.

Reason ______________________________ 1 mark

8 Jade does a survey to find out people's views on the following question.
'Did you learn anything from the lesson?'

a Say why each of the following would not be good responses to this question.

i Yes ☐ No ☐

Criticism ______________________________ 1 mark

ii A bit ☐ Don't know ☐ A lot ☐

Criticism ______________________________ 1 mark

iii Pythagoras ☐ Trigonometry ☐

Criticism ______________________________ 1 mark

b Say why this is a good response to the same question.

Mark on a scale from 1 (learnt a lot) to 5 (learnt little)

1 2 3 4 5

Reason ______________________________ 1 mark

9 There are 2000 students in a school.
To find out their views on vegetarianism some students do a survey. Melinda surveys 10 students picked at random from the school roll. Nandi surveys 30 students picked at random from the school roll. Owen surveys 100 students picked at random from the school roll.
Who will get the most reliable results? Give a reason for your answer.

Reason ______________________________ 1 mark

Box plots and cumulative frequency diagrams

level 8

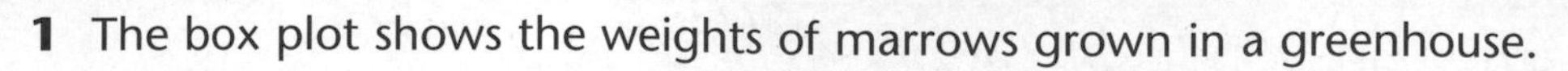

1 The box plot shows the weights of marrows grown in a greenhouse.

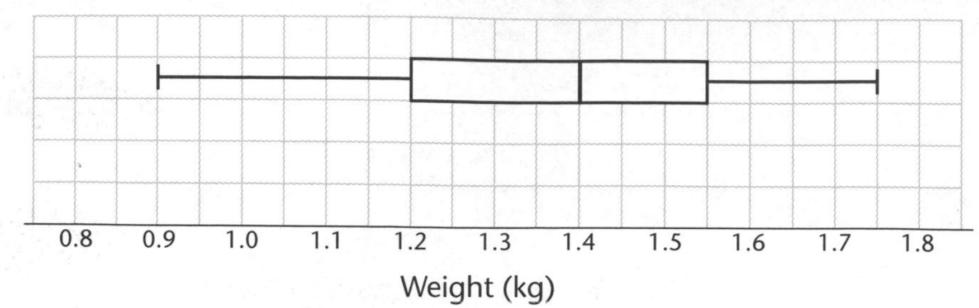

Below is some data about the weight of marrows grown outside.

Lowest value	Lower quartile	Median	Upper quartile	Highest value
0.8 kg	1.0 kg	1.3 kg	1.45 kg	1.6 kg

a On the same diagram draw the box plot for the weights of the marrows grown outside.

2 marks

b Comment on the difference between the average weight of the marrows grown outside and those grown inside.

____

1 mark

c Comment on the difference between the consistency of the weight of the marrows grown outside and those grown inside.

____

2 The box plot shows the marks (out of 10) for the boys in form 7B in a mathematics quiz.

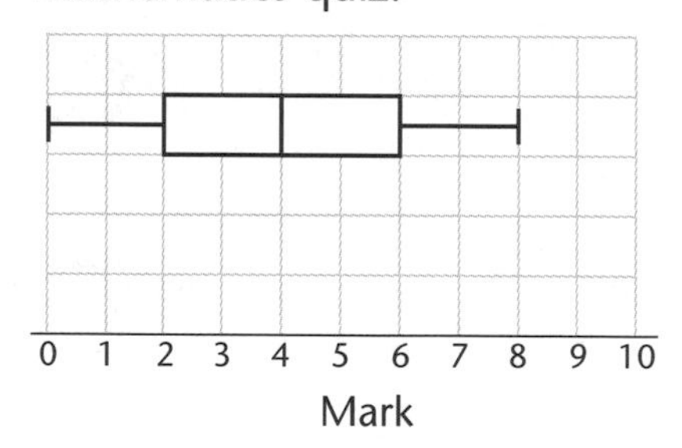

a The distribution of marks for the boys is symmetrical.
Explain how you can tell.

____

1 mark

b The following is information about the marks for the girls in 7B on the same quiz.

- The median mark was the same as the upper quartile for the boys.
- The lowest mark for the girls was the same as the lowest mark for the boys.
- The range of the girls' marks was 1 more than the range of the boys.
- The upper quartile for the girls was 1 more than their median.
- The interquartile range for the girls was the same as the interquartile range for the boys.

Use this information to draw a box plot for the girls on the same diagram.

2 marks

level 8

3 The cumulative frequency graph shows the heights of 100 girls.

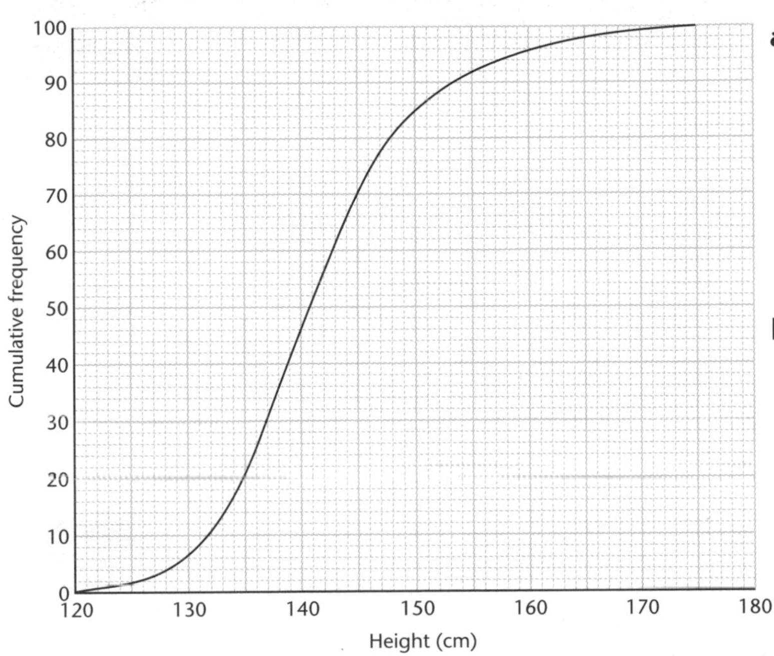

a Use the graph to estimate

i the median height

_______________ cm

ii the interquartile range

_______________ cm

b How many girls are over 155 cm?

____

4 The cumulative frequency graph shows the diameters of 60 oak trees in a wood.
The box plot shows the diameters of 60 oak trees in a parkland.

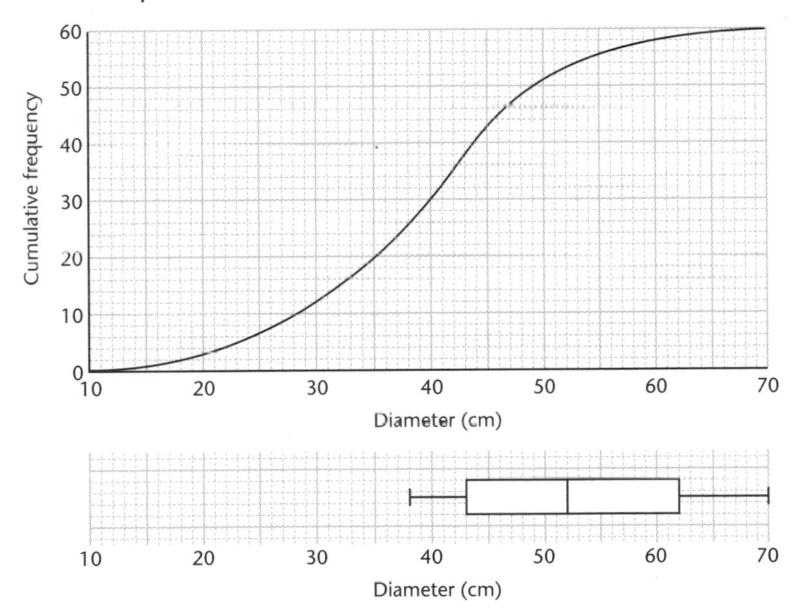

a What is the interquartile range and median of the oak trees in the wood?

interquartile range _______________ median _______________

b What is the interquartile range and median of the oak trees in the parkland?

interquartile range _______________ median _______________

c Comment on the differences between the distributions of the diameters of the trees in the wood and the parkland.

____

____

HANDLING DATA Probability 1

level 6

1 In a youth club, the probability that a member picked at random is a girl is $\frac{4}{7}$.
What is the probability that a member picked at random is a boy?

____

1 mark

2 A bag contains 1 blue and 4 red balls.
Some blue balls are to be added to the bag to make the chance of picking a blue ball at random $\frac{1}{2}$.
How many blue balls should be added? ________________

1 mark

3 A box of toffees contains nut and plain. The probability of getting a plain toffee is $\frac{9}{20}$. What is the probability of getting a nut toffee?

____

1 mark

4 Here are four events.

A Throwing a three with a dice.

B Picking a vowel at random from the letters:

D I S T R I B U T I O N

C The next person that comes into the room has a birthday in January.

D Throwing a number that is a factor of 24 with an ordinary dice.

Mark each event on the following probability scale.

0 $\frac{1}{2}$ 1

2 marks

5 This bag contains 4 white balls, 6 black balls and 5 striped balls.
A ball is taken from the bag at random.

a What is the probability it is black? ________________

1 mark

b What is the probability it is not striped? ________________

1 mark

level 6

6 The following cards are placed face down and shuffled.

S T A T I S T I C S

a A card is picked at random. What is the probability it is **not** a letter S or a letter T?

_______________ 1 mark

b A card is picked at random. It is a vowel. **It is thrown away**.
Another card is picked at random.
What is the probability it is a letter **S or T**?

_______________ 1 mark

7 A bag contains 10 coloured balls.
A ball is taken out, its colour noted and then replaced.
This is repeated 1000 times.
The results are red 822 times, blue 178 times.
How many **a** red balls **b** blue balls are in the bag?

Red _______________ Blue _______________ 2 marks

8 The sample space diagram shows the outcomes from throwing two coins.
There are four outcomes altogether.

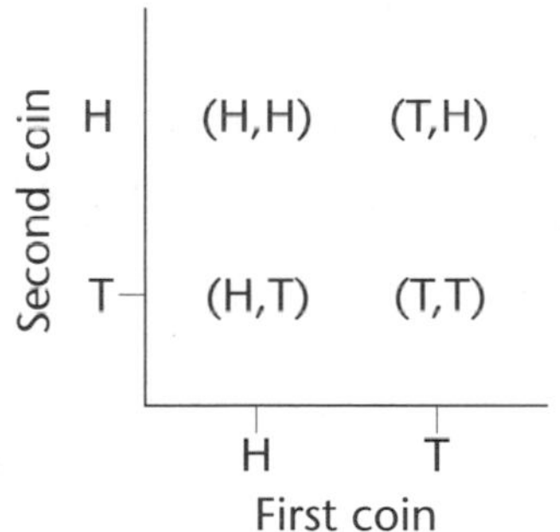

a What is the probability of throwing two heads with two coins?

_______________ 1 mark

b What is the probability of throwing a head and a tail in any order with two coins?

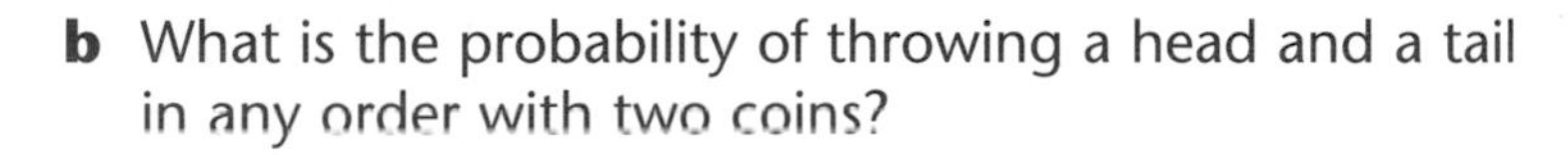

_______________ 1 mark

9 The sample space diagram shows the outcomes for throwing two dice.

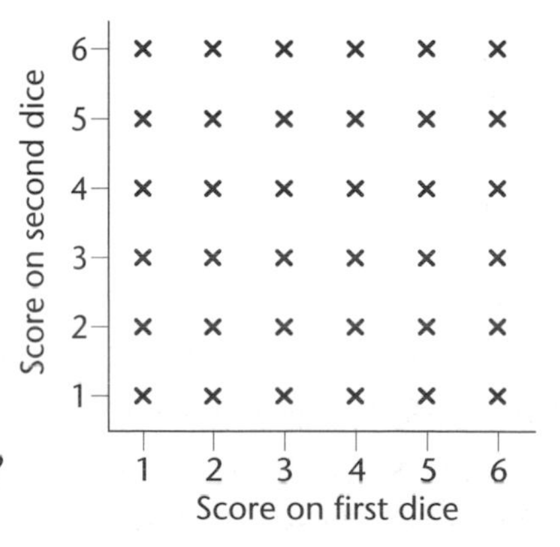

a What is the probability of throwing a 'double', i.e. the same score on each dice?

_______________ 1 mark

b What is the probability of a score of 4?

_______________ 1 mark

c What is the probability of a score of 10 or over?

_______________ 1 mark

Probability 2

level 8

1 A bag contains 1 blue and 4 red balls.
A ball is taken from the bag and **not replaced**.
Another ball is then taken out at random.

a If the first ball taken out is blue, what is the probability that the second ball taken out is

i blue? __________ 1 mark

ii red? __________ 1 mark

b If the first ball taken out is red, what is the probability that the second ball taken out is

i blue? __________ 1 mark

ii red? __________ 1 mark

2 This bag contains 4 white balls and 6 black balls.
A ball is taken from the bag at random and then replaced.
Another ball is then taken from the bag.

a Complete the tree diagram.

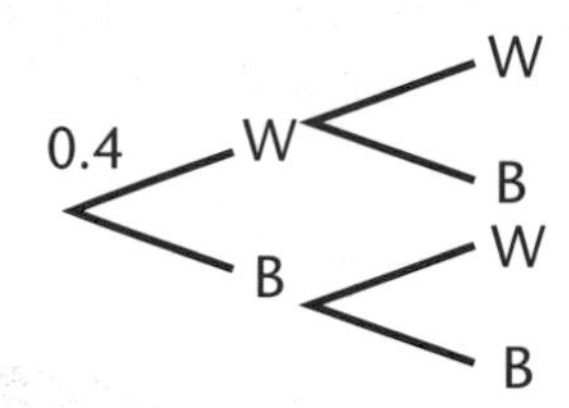

1 mark

b What is the probability that the balls are

i both white? __________

2 marks

ii both black? __________ 2 marks

3 This bag contains 4 white balls and 6 black balls.
A ball is taken from the bag at random and **not** replaced.
Another ball is then taken out.

a What is the probability that both balls are black?

__________ 2 marks

b What is the probability that both balls are the same colour?

__________ 2 marks

4 Two bags contain red and blue balls. Bag A has 5 red and 4 blue balls.
Bag B has 4 red and 3 blue balls. I can choose a ball from either bag.
I want to choose a red ball. Which bag should I choose?

__________ 2 marks

level 8

5 A company makes pens. The company knows that the probability that any pen will be defective is $\frac{1}{25}$.

a A box contains 1000 pens. How many of the pens are likely to be defective?

______________________ 1 mark

b Two pens are tested at random. Calculate the probability that both pens are defective.

______________________ 2 marks

c Calculate the probability that only one of the two pens will be defective.

______________________ 2 marks

6 **a** A fair dice is thrown once. What is the probability of scoring

i 2? ______________________

ii any value other than 2? ______________________ 2 marks

b A game costs 10p a go. A fair dice is thrown twice.

If it lands on a 2 twice then the player wins £1. If it lands on 2 once in the two throws the player gets their 10p back. If it does not land on 2 in either throw the player loses their money.

i Show that the probability of getting your money back is $\frac{10}{36}$

______________________ 2 marks

ii Simon has 36 goes. How much can he expect to win or lose?

______________________ 2 marks

7 A bag contains 20 balls. The balls are either white or black. Four people take balls out and replace them a different number of times.

Person	Pete	Rose	Sue	Tom
Number of goes	10	50	100	1000
White balls	3	18	33	320
Black balls	7	32	67	680

a Which person's results are the most reliable? Give a reason why.

______________________ 1 mark

b Estimate how many balls of each colour there are in the bag. Justify your answer.

____

______________________ 2 marks

Practice Paper 1

Time allowed 60 minutes.
You may not use a calculator on this paper.

---

1 This quadrilateral has one acute angle and three obtuse angles.

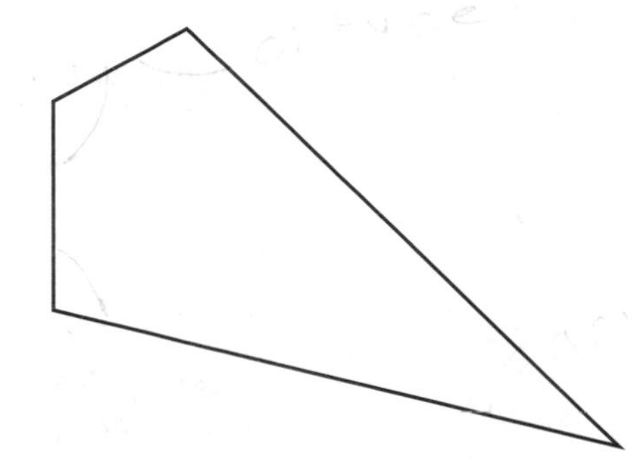

a Now draw a quadrilateral which has two acute angles and two obtuse angles.

1 mark

b Explain why you cannot draw a quadrilateral which has four acute angles.

____

1 mark

2 Anna buys a box of chocolates that are all the same size and shape. The box contains 12 milk chocolates, 8 plain chocolates and 5 white chocolates. Anna takes a chocolate from the box at random.

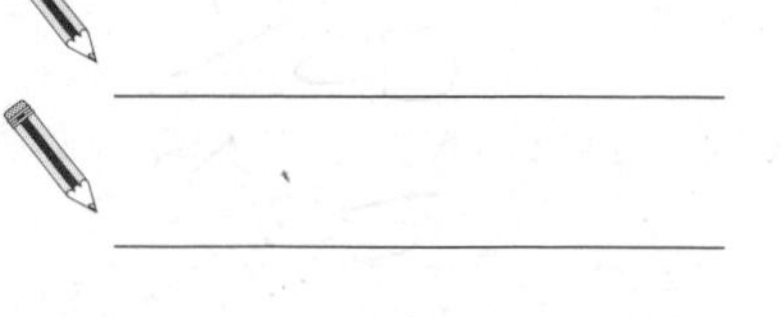

a What is the probability that she takes a milk chocolate? __________

1 mark

b What is the probability that she takes a plain chocolate? __________

1 mark

c What is the probability that she does not take a white chocolate? __________

1 mark

3 A single ticket on the metro costs £1.35.
Dave buys a book of 25 single tickets, which costs him £30.
How much does Dave save by buying a book of tickets?

£ ______________________ *2 marks*

4 When $a = 6$, $b = 5$ and $c = 2$,

a work out the value of the following:

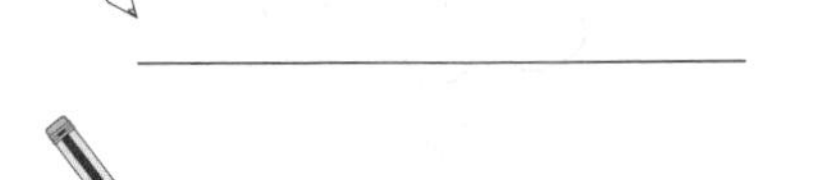

$a + 2b + c$ ______________________ *1 mark*

$3a + b - 2c$ ______________________ *1 mark*

b If $a + b + c + d = 20$, work out the value of d.

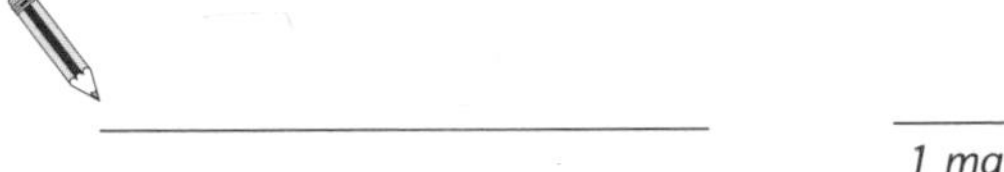

______________________ *1 mark*

5 **a** Complete the following sentences.

__________ out of 200 is the same as 40% *1 mark*

30 out of 50 is the same as __________% *1 mark*

b __________ out of __________ is the same as 5% *1 mark*

6 Here is a fraction strip.

<table>
<tr><td colspan="6">$\frac{1}{2}$</td><td colspan="6"></td></tr>
<tr><td colspan="4">$\frac{1}{3}$</td><td colspan="4"></td><td colspan="4"></td></tr>
<tr><td colspan="3">$\frac{1}{4}$</td><td colspan="3"></td><td colspan="3"></td><td colspan="3"></td></tr>
<tr><td>$\frac{1}{12}$</td><td></td><td></td><td></td><td></td><td></td><td></td><td></td><td></td><td></td><td></td><td></td></tr>
</table>

Use the fraction strip to help you work out the following:

$\frac{1}{2} + \frac{5}{12} =$ ______________________ *1 mark*

$\frac{1}{4} + \frac{1}{3} =$ ______________________ *1 mark*

$\frac{3}{4} - \frac{5}{12} =$ ______________________ *1 mark*

7 **a** Complete the table for the mapping $y = x + 5$.

x	2	4	6
y	7		

1 mark

b Complete the table for the mapping $y = 2x - 3$.

x	2	4	6
y	1		

1 mark

c Write down the mapping for this table.

x	2	4	6
y	2	3	4

$y =$ ____________

1 mark

8 Here are three cuboids.

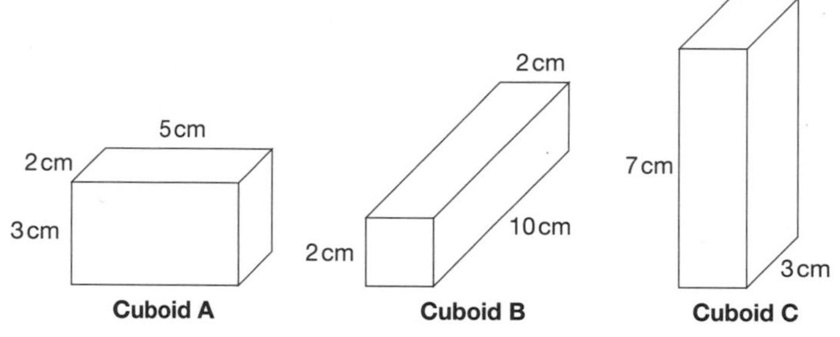

a Which of the cuboids has the largest surface area?
Tick (✓) the correct box.

Cuboid A ☐ Cuboid B ☐ Cuboid C ☐

2 marks

b Which of the cuboids has the largest volume?
Tick (✓) the correct box.

Cuboid A ☐ Cuboid B ☐ Cuboid C ☐

2 marks

c Cuboid D has the same volume as Cuboid A.
Cuboid D has a length of 10 cm and width of 3 cm.
What is its height?

____________ cm

1 mark

9 Three quadrilaterals are drawn on square grids below.

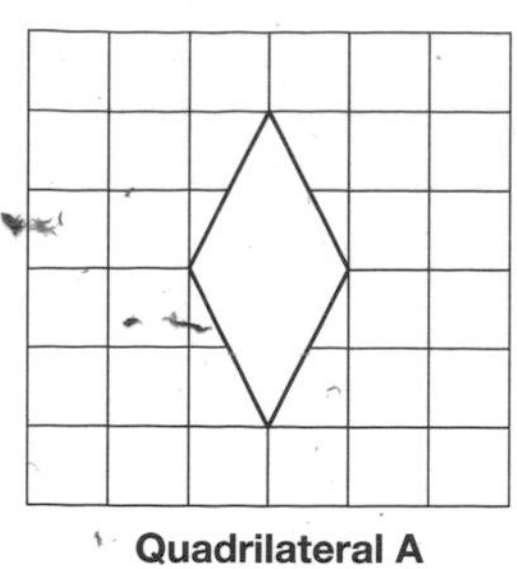
Quadrilateral A

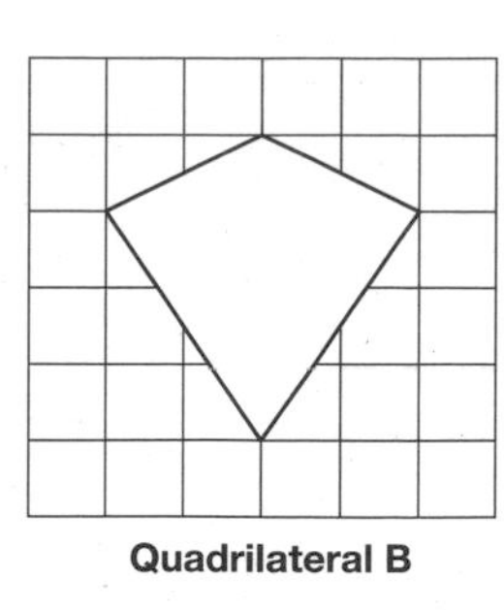
Quadrilateral B

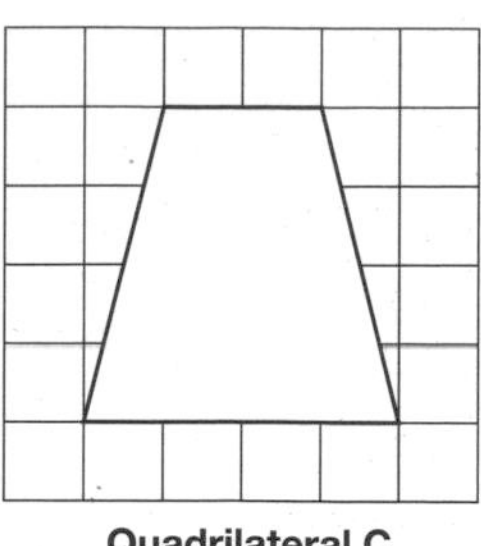
Quadrilateral C

a Is Quadrilateral A a square? Tick (✓) the correct box. 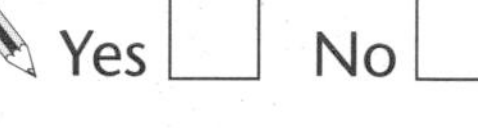 Yes ☐ No ☐
Explain your answer.

____

1 mark

b Is Quadrilateral B a kite? Tick (✓) the correct box. Yes ☐ No ☐
Explain your answer.

____

1 mark

c Is Quadrilateral C a parallelogram? Tick (✓) the correct box. Yes ☐ No ☐
Explain your answer.

____

1 mark

10 Fill in the missing numbers in the boxes.

12 + ☐ = 10

1 mark

6 – ☐ = 10

1 mark

–2 x ☐ = 10

1 mark

11 Work out $\frac{2}{3} \times \frac{3}{8}$

Write your answer as a fraction in its simplest form. ____________

2 marks

12

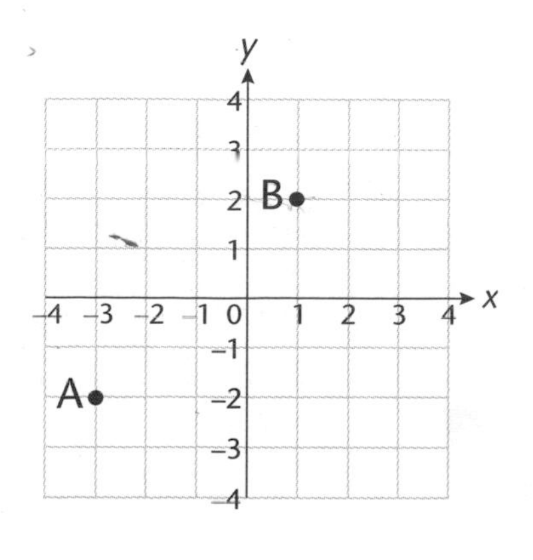

a What are the coordinates of the point A?

(________ , ________)

1 mark

b What are the coordinates of the mid-point of AB?

(________ , ________)

1 mark

13 Solve the following equations.

$2x + 3 = 11$ $\qquad x =$ ______ *1 mark*

$3(y - 2) = 9$ $\qquad y =$ ______ *1 mark*

$3z - 4 = z + 2$ $\qquad z =$ ______ *1 mark*

14 Rearrange the following equations.

$x + y = 7$ $\qquad x =$ ______ *1 mark*

$3w = z$ $\qquad w =$ ______ *1 mark*

15 Multiply out the expression $(x + 3)(x - 4)$.
Write your answer as simply as possible.

______ *2 marks*

16 The scatter graph shows the arm span and height for 10 students.

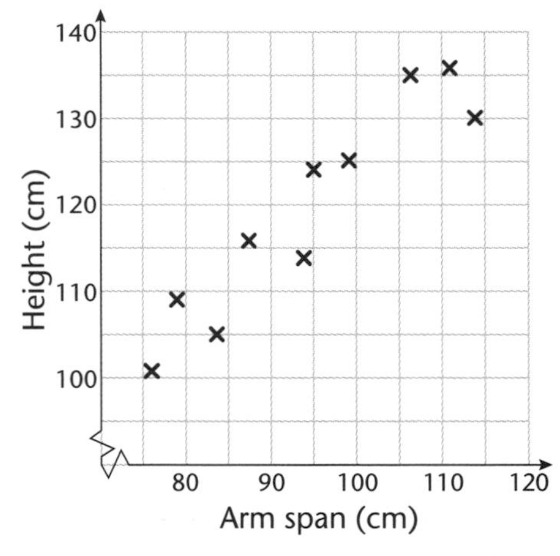

a Describe the relationship between arm span and height.

______ *1 mark*

b Draw a line of best fit on the scatter graph. *1 mark*

c A student grows 10 cm in height in a year.
By how much would you expect the arm span to increase?

______ cm *1 mark*

d A baby has a height at birth of 35 cm. Can you use the scatter graph to predict the baby's arm span?
Explain your answer.

______ *1 mark*

17 15 members of a slimming club record their weights before and after dieting for 3 months. The stem-and-leaf diagrams show the weights before and after the diet.

Before

Stem	Leaves
5	
6	5 7 9
7	2 3 6 6 6 9
8	0 2 4 8
9	2 5

After

Stem	Leaves
5	4 4 6 9
6	3 5 6 8 4
7	1 2 2 2 9
8	3
9	

Key:
7 | 2 means 72 kg

Complete the following sentences.

a Before the diet the heaviest person was ________ kg and ________ members of the club were over 70 kg.

1 mark

b After the diet the heaviest person was ________ kg and ________ members of the club were over 70 kg.

1 mark

c Before the diet the modal weight was ________ kg and the range of the weights was ________ kg.

1 mark

18 a Work out 27 x 33.

____

2 marks

b A 2 digit number is multiplied by another 2 digit number.
What is the least number of digits the answer could have?
Explain your answer.

____

1 mark

What is the greatest number of digits the answer could have?
Explain your answer.

____

1 mark

19 A box contains some black and white counters.
There are 10 counters in the box altogether.
Amy and Zara do an experiment to find out how many counters of each colour there are in the box.
They each take a counter at random, record the colour and replace the counter.
The table shows their results.

Pupil	Number of trials	Number of black counters	Number of white counters
Amy	100	36	64
Zara	500	210	290

a Using Amy's results, obtain an estimate of the probability of taking a black counter from the box.

____

1 mark

b Using Zara's results, obtain an estimate of the probability of taking a white counter from the box.

____

1 mark

c Which of the two probabilities in **a** and **b** is likely to be the more accurate? Give a reason for your answer.

____

1 mark

20 Work out the following.
Give your answers in standard form.

$1.2 \times 10^4 \times 2 \times 10^2 =$ _______________

1 mark

$1.2 \times 10^8 \div 2 \times 10^2 =$ _______________

1 mark

21 a and b are connected by the relationship $a^b = 64$.
One possible pair of values is $a = 8$, $b = 2$.
Give 3 other possible pairs of values for a and b that satisfy the relationship.

$a =$ _______________ $b =$ _______________

$a =$ _______________ $b =$ _______________

$a =$ _______________ $b =$ _______________

3 marks

Practice Paper 2

Time allowed 60 minutes.
You may use a calculator on this paper.

---

1 The pie chart shows the replies to a survey on holiday destinations.

Britain | America | 84° | Europe | Other

a 7 people answered 'America'.
How many people were in the survey altogether? ______________ people

2 marks

b A different survey of 20 people were asked if they preferred staying in Britain or going abroad for their holidays.
9 people said they preferred to stay in Britain.
On a pie chart, what would the angle be for 'Staying in Britain'?

______________ degrees

2 marks

2 **a** For each number in the table write a factor of that number that is **between** 10 and 20.

Number	Factor between 10 and 20
48	
150	
51	

3 marks

b Is 150 a multiple of 60? Tick (✓) Yes or No.

Yes ☐ No ☐

Explain how you know.

____

1 mark

3 Here are eight number cards.

−3	−1	−1	0	2	6	8	9

a What is the range of the numbers? ____________ *1 mark*

b What is the sum of the numbers? ____________ *1 mark*

c What is the mode of the numbers? ____________ *1 mark*

d What is the median of the numbers? ____________ *1 mark*

e What is the mean of the numbers? ____________ *2 marks*

4 Here is part of a number grid.

23	24	25	26	27	28
33	34	35	36	37	38
43	44	45	46	47	48

From these numbers, write down one that is:

a a prime number ____________ *1 mark*

b a square number ____________ *1 mark*

c Explain why a square number could never be a prime number.

____ *1 mark*

5 a ABC is an isosceles triangle.

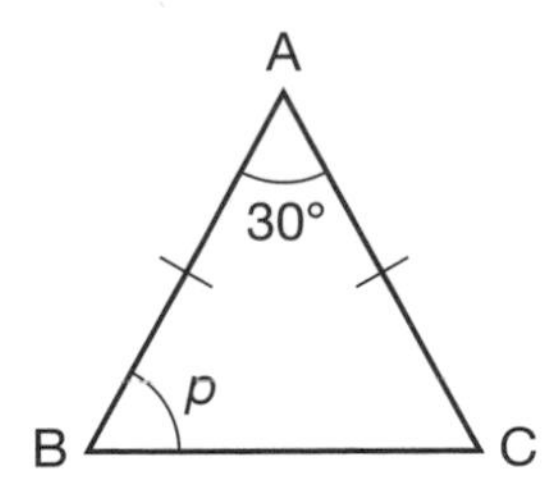

What is angle p?

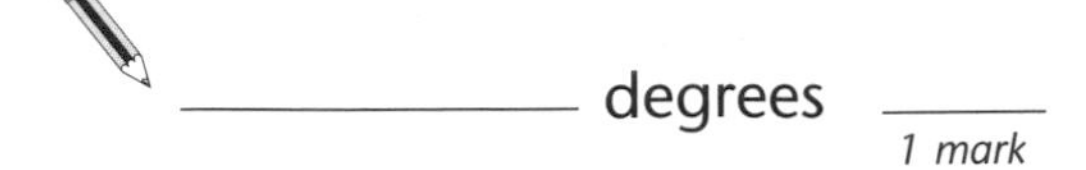

b This diagram is not drawn accurately.
Calculate the size of angle m.
Show your working.

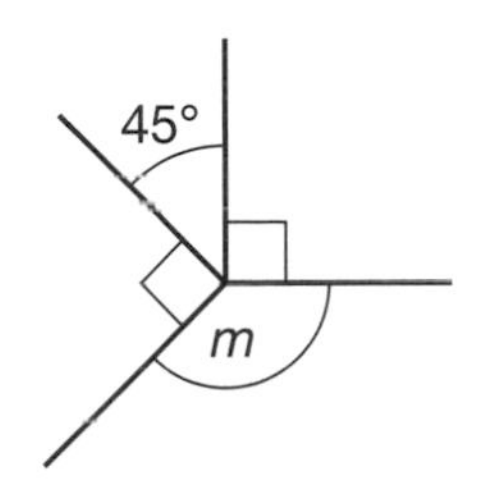

_________ degrees

1 mark

6 The triangle ABC below is drawn accurately.

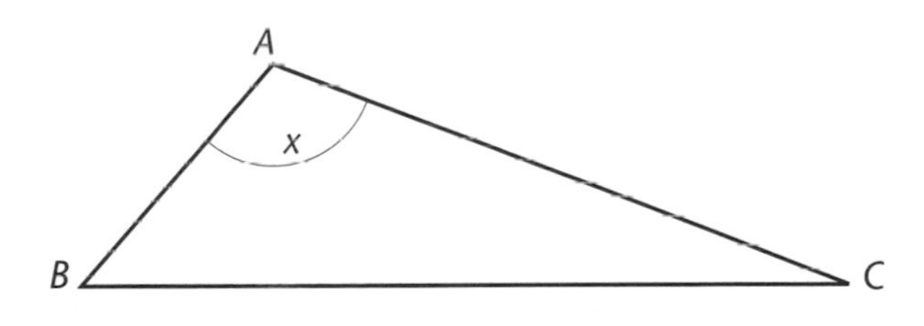

a Measure accurately the angle marked x.

_________ degrees

1 mark

b The drawing is a scale drawing of a building plot.
The scale is **1 cm represents 50 metres**.
What is the actual length represented by BC on the diagram?

_________ metres

2 marks

7 A 50p coin has a mass of 8 grams.
How much is one kilogram of 50p coins worth?

£ _________

3 marks

8 The graph shows a straight line.

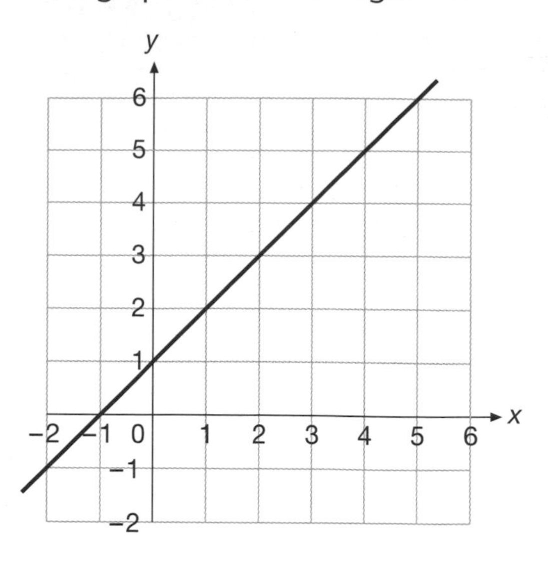

a Fill in the table with some of the points on the line.

(x, y)	(___ , ___)	(___ , ___)	(___ , ___)

2 marks

b Write down the equation of the line. ______________________

1 mark

c On the graph draw the line $y = x + 3$

1 mark

9 A bicycle wheel has a diameter of 70 cm.

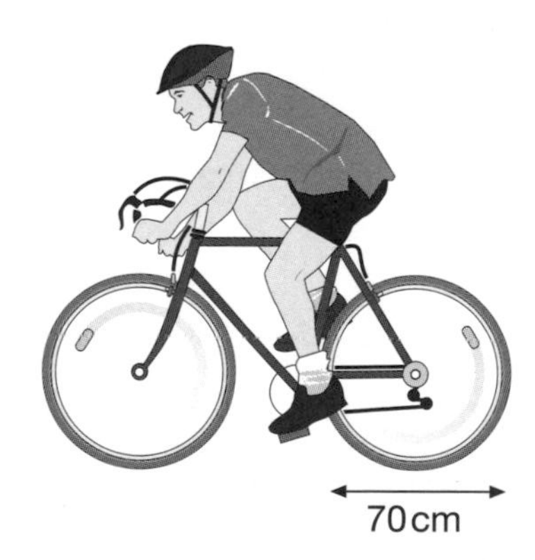

a What is the circumference of the wheel?

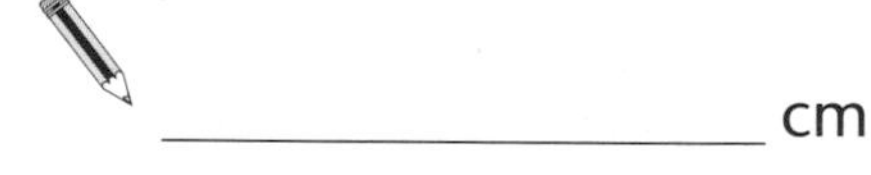

______________________ cm

1 mark

b During a 5 kilometre race, approximately how many times will the wheel turn?

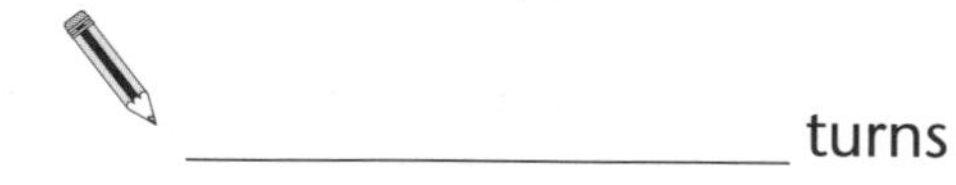

______________________ turns

2 marks

10 Use your calculator to work out

$(52 + 25) \times (41 - 19) =$ ______________________

1 mark

$\dfrac{52 + 25}{41 - 19} =$ ______________________

1 mark

11 The standard measure for different paper sizes are A1, A2, A3 etc....
The standard measure for envelopes are C1, C2, C3 etc...
All paper and envelope sizes have the width and height in the same ratio of approximately 1 : 1.4

Ratio of width to height is 1 : 1.4

a Work out the height of a piece of A4 paper that is 21 cm wide.

A4 paper has a height of

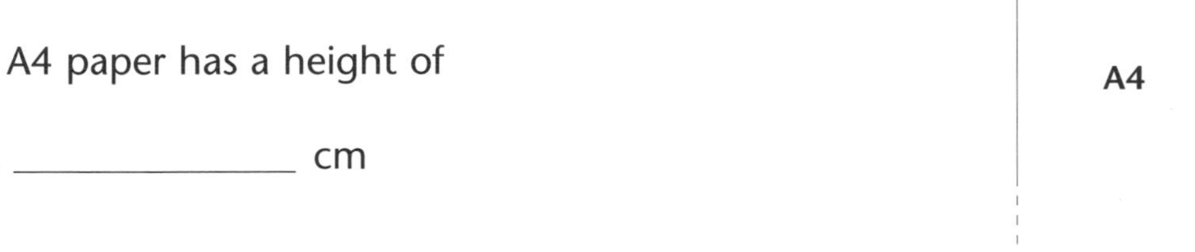

__________ cm

1 mark

b Work out the width of a C5 envelope that is 229 mm high.
Give your answer to the nearest millimetre.

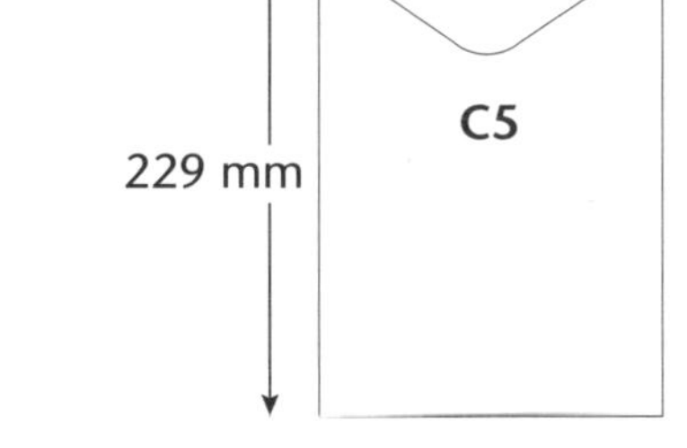

A C5 envelope has a width of

__________ mm

1 mark

c Will an A4 piece of paper, when folded in half, fit inside a C5 envelope?
Explain your answer.

____

____

1 mark

12 Some information about the capacity of two football grounds is shown in the table.

	Manchester United	Manchester City
Total capacity	67 500	48 000
Percentage of executive seats	3.4%	4.9%

Which club has the most executive seats and by how many? __________

2 marks

13 Look at this equation.

$3(2x + 13) = 76 + 4x$

Is $x = 18.5$ a solution of this equation? Tick [✓] Yes or No.

Yes ☐ No ☐

Explain your answer. ______________________________

1 mark

14 Write these expressions as simply as possible.

$7 - 3a + 9 - 7a =$ ______________ *1 mark*

$b^2 + 6b - 2b =$ ______________ *1 mark*

$6c \times 4c =$ ______________ *1 mark*

$\frac{12d^3}{3d} =$ ______________ *1 mark*

15 The table shows the number of children in 20 families.

Number of children per family	Frequency
0	2
1	5
2	8
3	4
4	0
5	1

Calculate the mean number of children for the 20 families.

2 marks

16 The star is formed using a regular octagon.

a Work out the value of angle a.

b Work out the value of angle b.

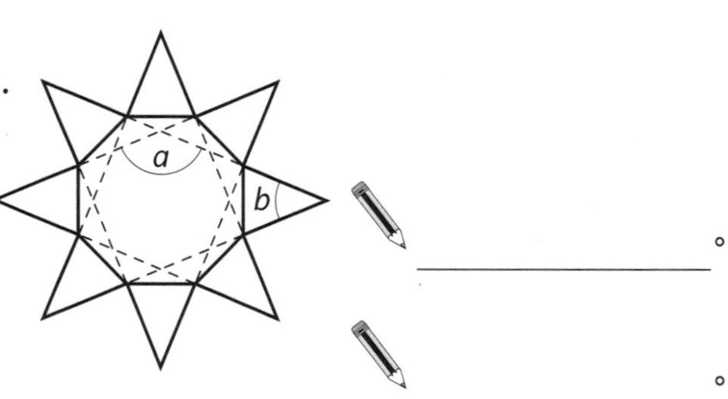

1 mark

1 mark

17 Construct an angle of 60° at the point *A*.
Show your construction lines and arcs.

A •

2 marks

18 Calculate the length x in the following triangles.
Show your working clearly.
Give your answers to 3 significant figures.

a

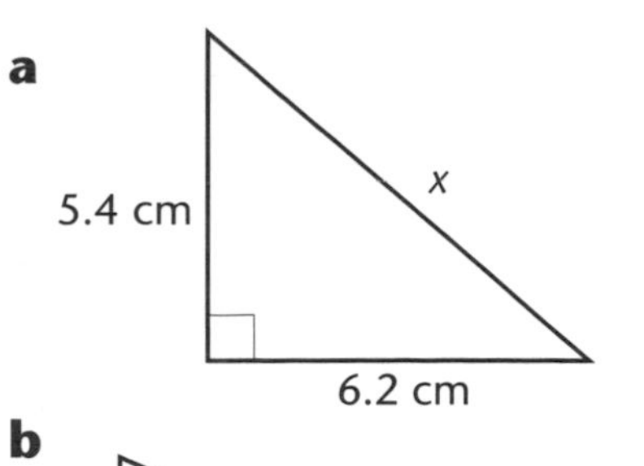

b

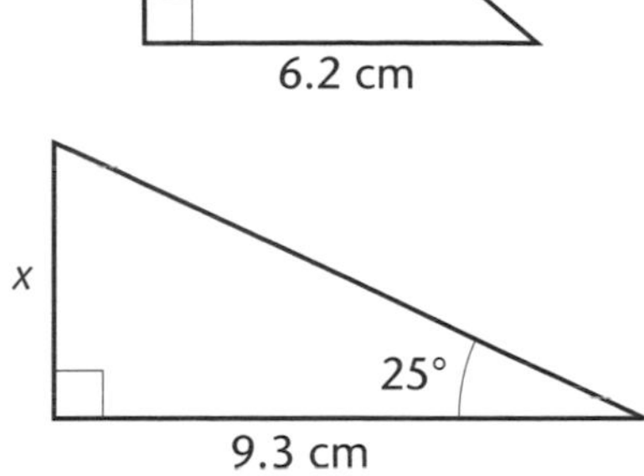

19 The length of a rectangle is decreased by 10%.
The width of the rectangle is increased by 20%.
By what percentage does the area increase or decrease?

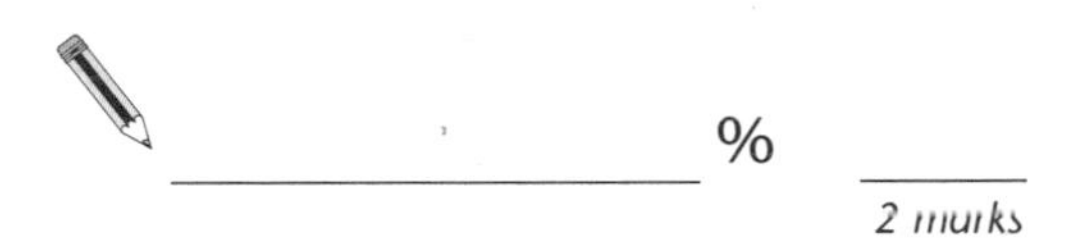

20 What is the nth term of the series
4, 7, 10, 13, 16, ...

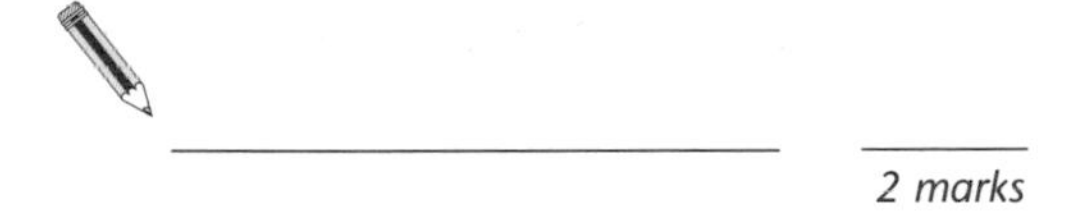

21 The diagram shows a rectangle inside a right-angled triangle.

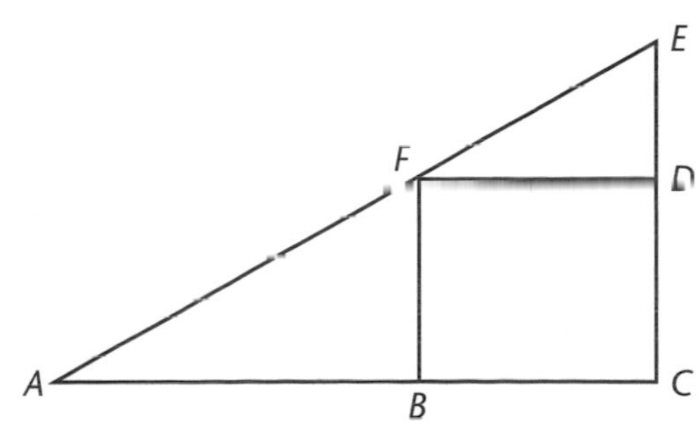

$ED = 4$ cm, $CD = 6$ cm, $EF = 8$ cm.
Find the length AE.

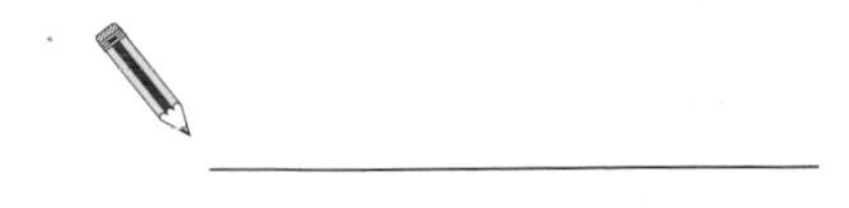

2 marks

Mental Mathematics Test

Here is an example of a mental mathematics test. Ask someone to read out the questions and write your answers in the book. For the first group of questions you will have 5 seconds to work out each answer and write it down.

Time: 5 seconds

1 Multiply four point three by ten.

2 How many metres are in 300 centimetres?

3 What is two-fifths of thirty-five?

4 Subtract four from minus six.

5 Look at the equation. When x equals six, what is the value of y?

6 What is four point five divided by five?

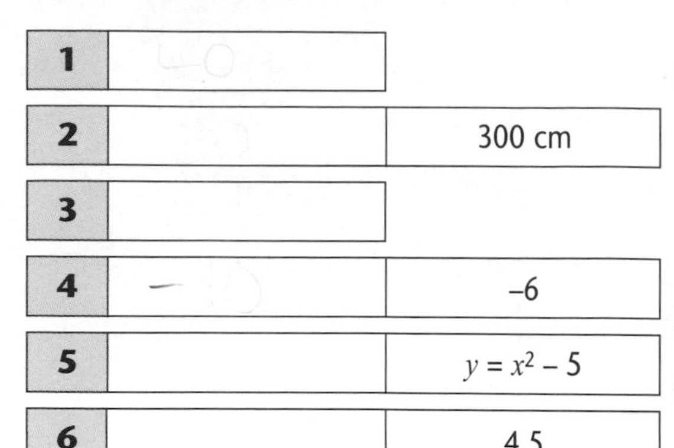

7 To the nearest ten kilometres the length of a motorway is ninety kilometres.

What is the greatest value the length of the motorway could be?

7	km

For the next group of questions you will have 10 seconds to work out each answer and write it down.

Time: 10 seconds

8 The chart shows the number of hours of TV watched by a child in a week.

On which day was 3 hours of TV watched?

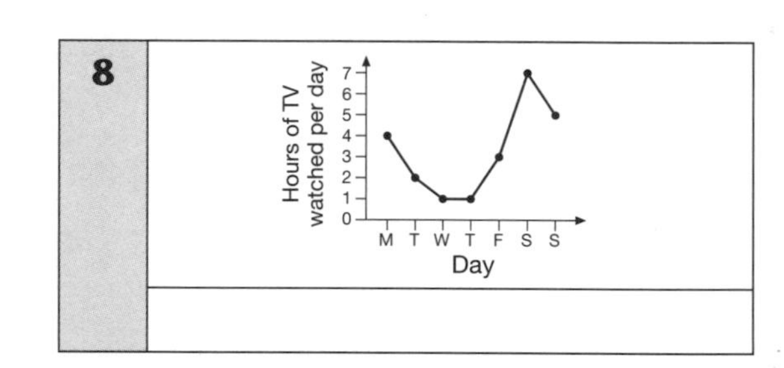

9 A robot moves so that it is always the same distance from a fixed point.

What is the name of the shape of the robot's path?

9	

10 Look at the grid. Write down the coordinates of the mid-point of AB.

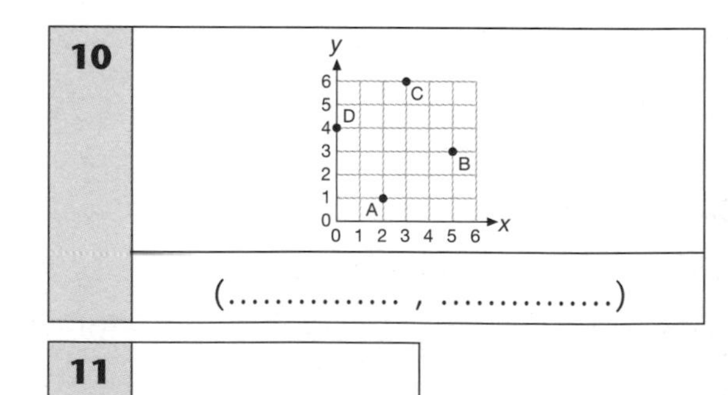

11 How many quarters are there in two and a half?

11	

12 Think about the mass two kilograms.

About how many pounds is that?

Circle the best answer on the answer sheet.

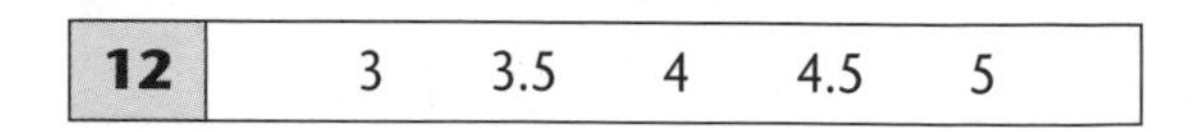

13 Look at the fraction.

Write it in its simplest form.

13		$\frac{85}{100}$

14 In a survey one-third of the people asked preferred to go abroad for their holidays.

What percentage is this?

14	

15 What is the area of this rectangle?

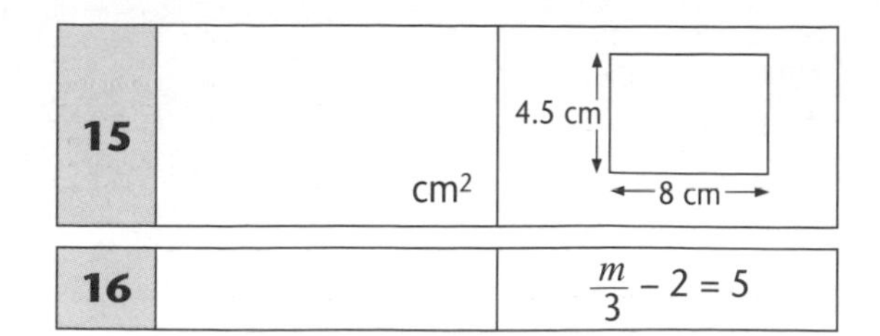

16 Look at the equation. Solve it to find the value of m.

16 $\frac{m}{3} - 2 = 5$

17 The average weight of a male squirrel is 500 grams.

Female squirrels have an average weight that is 5% less than this.

What is the average weight of a female squirrel?

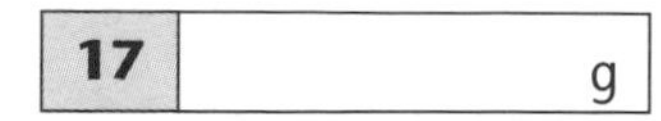

18 A cardboard box measures half a metre by thirty centimetres by twenty centimetres.

Which of the calculations on the answer sheet will give the volume of the box?

Ring the correct answer.

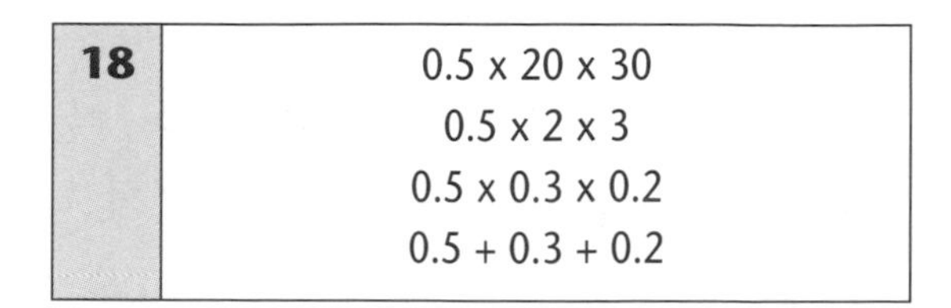

19 What is a quarter of two thirds of sixty?

20 Look at the inequality.

How many integer solutions are there?

20 $3 \leq n \leq 9$

For the next group of questions you will have 15 seconds to work out each answer and write it down.

Time: 15 seconds

21 Write down a factor of 48 that is bigger than ten but less than twenty.

22 The first odd number is one. What is the hundredth odd number?

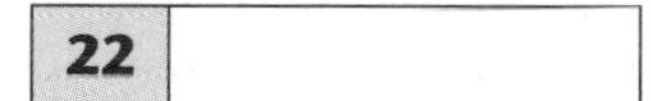

23 On the grid sketch the line $x + y = 4$.

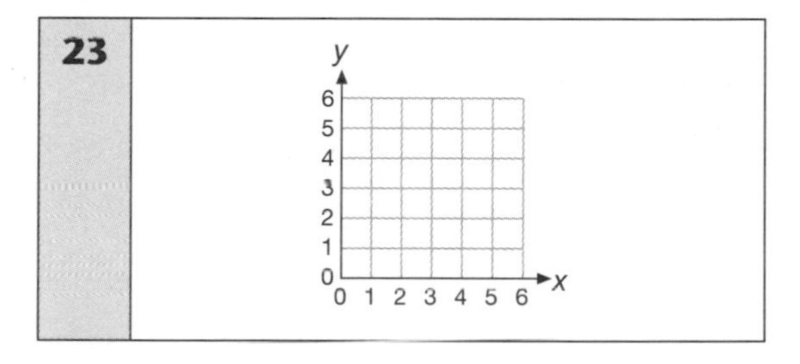

24 What is the area of a circle with a radius of 3 centimetres?

Give your answer in terms of π.

25 I can make twenty-four different four-digit numbers from the digits one, two, three and four.

How many of these will be odd numbers?

26 Look at the calculation.

Write down an approximate answer.

26 $\frac{38.5 \times 51.6}{4.89}$

27 Complete the factorisation.

27 $x^2 - 16 = (x + 4)$ (.....................)

28 A bag contains only red and blue balls.

There are twice as many blue balls as red balls.

I take a ball at random from the bag.

What is the probability that the ball will be red?

28

29 What 3-D shape has four vertices?

29

30 What is the product of all the integers from 1 to 5?

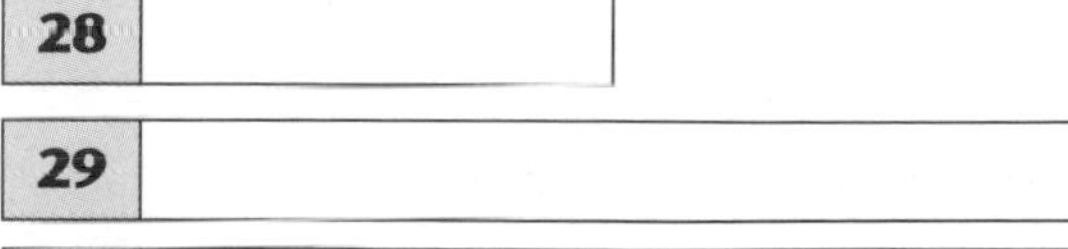

Revision checklist

Number

I am able to:

- Multiply and divide decimals
- Add, subtract, multiply and divide negative numbers
- Find a percentage of a quantity
- Find one quantity as a percentage of another
- Find the new value after a percentage increase or decrease
- Find the original amount after a percentage increase or decrease
- Divide a quantity in a given ratio
- Work out powers of numbers
- Round off numbers and give the limits of accuracy
- Write numbers in and calculate with standard form

Algebra

I am able to:

- Simplify algebraic expressions
- Use formulae expressed in words and in symbols
- Factorise simple expressions
- Expand and simplify a pair of linear brackets
- Draw a graph by plotting points
- Recognise graphs of the form $x = y$, $y = b$ and $y = x$; $y = x^2$, $y = x^3$ and $y = \frac{1}{y}$
- Interpret real-life graphs
- Solve simple equations by rearrangement
- Solve equations where the unknown appears on both sides of the equals sign
- Solve equations involving brackets
- Solve equations using trial and improvement
- Solve a pair of simultaneous equations

Shape, Space and Measures

I am able to:

- Calculate angles from known facts
- Plot and measure bearings
- Describe and draw shapes that have been transformed by enlargement, rotation and reflection
- Calculate the interior and exterior angles of regular polygons
- Recognise the standard 3-D shapes and their nets
- Calculate the perimeter and area of standard shapes
- Calculate the circumference and area of a circle
- Calculate the volume of cuboids
- Use a ruler and compasses to do constructions
- Find the loci of points obeying certain rules
- Use similar triangles to calculate sides of triangles
- Use Pythagoras' theorem to calculate sides of right-angled triangles
- Use trigonometry to calculate sides and angles of right-angled triangles
- Find the length of an arc and area of a sector
- Find angles using the circle theorems

Handling Data

I am able to:

- Draw and interpret pie charts
- Draw and interpret scatter diagrams and lines of best fit
- Design and criticise questions for surveys
- Compare distributions using statistical measures
- Use box plots and cumulative frequency diagrams to find medians and interquartile ranges
- Calculate the probability of an event
- Calculate the probability of a combined event from a sample space diagram
- Calculate the probability of a combined event from a tree diagram

Workbook answers

Pages 100–101 Multiplying and dividing decimals

1 a 3400 **b** 0.074

2 a 54.3 **b** 34.7 **c** 0.672 **d** 0.0807

3 a 0.5 **b** 2.3 **c** 0.006 **d** 100

4 a £68.40 **b** £11.90

5 £106.72 *(1 mark for £26.97 or £79.75)*

6 £65

7 a 32.2 **b** 5.9 **c** 41.6 **d** 8.3

8 £42.80

9 $\frac{1}{5}$ of £46 = £9.20, as $\frac{1}{8}$ of £73 = £9.125

Pages 102–103 Long multiplication and division

Questions worth 2 marks get full marks for a correct answer and 1 mark for correct working with one error. For example 27 x 32 = 20 x 32 + 7 x 32 = 640 + 214 = 754 would get 1 mark. (The correct answer is 640 + 224 = 764.)

1 a 864 **b** 7812 **c** 34 **d** 71

2 a 816 **b i** 20 **ii** 40

3 a 1196 **b** 19

4 a 360
b i 19
ii Yes, 19 x 15 = 285 so 13 left over

5 a £2200 *(1 mark for 14700)*
b £1300 *(1 mark for 11200)* **c** 15

6 19

7 30

Pages 104–105 Negative numbers

1 a –2 **b** –7 + –6 + –2 = –15
c –7 x 8 = –56

2 a 68 °F **b** –40 °F **c** –20 °C

3 b $<$ **c** = **d** =

4 a –11 **b** 11 **c** –4 **d** 42

5 a Any valid answer, e.g. –3 + –2
b Any valid answer, e.g. –7 – –2

6 a –14 **b** +7 **c** –7, 3

7 a –2 and –6 **b** –7

8 a –1 **b** –4

Pages 106–107 Adding and subtracting fractions

1 a 17 **b** 10

2 a $3\frac{3}{4}$ **b** $\frac{20}{7}$

3 a $\frac{11}{15}$ *(1 mark for $\frac{5}{15}$ or $\frac{6}{15}$)*
b $\frac{7}{20}$ *(1 mark for $\frac{12}{20}$ or $\frac{5}{20}$)*
c $\frac{39}{40}$ *(1 mark for $\frac{15}{14}$ or $\frac{24}{40}$)*
d $\frac{5}{24}$ *(1 mark for $\frac{21}{24}$ or $\frac{16}{24}$)*

4 a $5\frac{9}{20}$ *(1 mark for $\frac{9}{4} + \frac{16}{5}$)*
b $4\frac{21}{40}$ *(1 mark for $\frac{17}{5} + \frac{9}{8}$)*
c $\frac{9}{20}$ *(1 mark for $\frac{21}{5} - \frac{15}{4}$)*
d $1\frac{5}{12}$ *(1 mark for $\frac{11}{4} - \frac{4}{3}$)*

5 a $\frac{5}{12}$ *(1 mark for $\frac{20}{48}$)*
b $\frac{7}{12}$ *(1 mark for $\frac{28}{48}$)*

6 a $\frac{4}{25}$ **b** $\frac{2}{7}$ **c** $\frac{6}{25}$ *(1 mark for $\frac{2}{3} \times \frac{9}{25}$)*

Pages 108–109 Multiplying and dividing fractions

1 $\frac{1}{8}$

2 $\frac{3}{20}$

3 12

4 a $4\frac{1}{5}$ **b** $\frac{19}{6}$

5 a $\frac{1}{5}$ **b** $1\frac{1}{4}$ **c** $\frac{2}{7}$ **d** $\frac{2}{3}$

6 a $2\frac{7}{10}$ **b** $\frac{2}{3}$ **c** $8\frac{1}{4}$ **d** $\frac{9}{10}$

7 $3\frac{9}{20}$ cm² *(1 mark for $\frac{69}{20}$)*

8 a $1\frac{3}{8}$ cm² *(1 mark for $2\frac{3}{4}$)*
b $\frac{17}{24}$ cm² *(1 mark for $1\frac{5}{12}$)*

9 30 *(1 mark for $\frac{66}{1} \times \frac{5}{11}$)*

10 $1\frac{1}{4}$ cm *(1 mark for $\frac{8}{1} \times \frac{5}{32}$)*

Pages 110–111 Percentages 1

1 a 14% **b** Fish cakes **c** Salad
d 6 out of 20 (30%) is more than 6 out of 30 (20%) **e** Fish cakes (10% of each)

2 a £48 **b** £25.60 **c** £72 (10% off £80)

3 a South America **b** 13.3%

4 a 12.5% **b** 880 acres
c Increase. Crops will be 50%, cattle will be $33\frac{1}{3}$ % so fallow land will be $16\frac{2}{3}$%. *(1 mark for 330 ÷ 3 = 110)*

Pages 112–113 Percentages 2

1 45%

2 a 15 **b** 18

3 a 16% **b** 12.5%

4 6 out of 10 = 60%; 40 out of 64 = 62.5%; 32 out of 50 = 64%; 13 out of 20 = 65% *(1 mark for 3 correct)*

5 15% *(1 mark for $\frac{66}{440}$)*

6 9748 *(1 mark for 0.88^{10})*

7 a 1504 **b** 975

8 £350 *(1 mark for 308 ÷ 0.88)*

9 80 g *(1 mark for 92 ÷ 1.15)*

10 32% *(1 mark for 1.1 x 1.2 = 1.32)*

Pages 114–115 Ratio

1 a 2 : 3 **b** 3 : 5

2 800 g *(1 mark for 200 g)*

3 Any seven squares shaded

4 a £18 : £72 *(1 mark for 90 ÷ 5)*
b 60 kg : 90 kg *(1 mark for 150 ÷ 5)*

5 a 175 ml *(1 mark for 25 ml)*
b 100 ml *(1 mark for 50 ml)*

6 a 7 : 3 *(1 mark for 28 : 12)*
b 13 : 3 *(1 mark for 39 : 9)*

7 a Nadia £54.75 Naseem £91.25
b Nadia £113.15 Naseem £178.85

8 a 3 : 5 **b** 7 : 12
c Small bottle *(1 mark for 3 ÷ 7 or 5 ÷ 12)*

Pages 116–117 Powers and roots

1 a 64 **b** 2187

2 a 4^7 **b** 5^6

3 25^0, $\sqrt[3]{64}$, $\sqrt{25}$, 2^3, 3^2

4 a 3^3 **b** 4^2 and 2^4

5 a 1 (or 0) **b** Any number between 0 and 1

6 a 3 and 4 **b** 2 and 3

7 $a = 6, b = 3, c = 2$

8 $m = 2, n = 3$

9 a 5 cm **b** 150 cm^2

10 Any that work, e.g. $\sqrt{2+3} \approx 2.2$, $\sqrt{3} + \sqrt{2} \approx 3.1$

11 Any that work, e.g. $\sqrt{(4 \times 9)} = \sqrt{36} = 6$, $\sqrt{4} \times \sqrt{9} = 2 \times 3 = 6$

12 $a = 4, b = 2$ (or vice versa)

Pages 118–119 Approximations and limits

1 a 50 **b** 0.007

2

Number	Rounded to 1 s.f.	Rounded to 2 s.f.	Rounded to 3 s.f.
5.682	6	5.7	5.68
34 639	30 000	35 000	34 600
0.09938	0.1	0.099	0.0994

3 a 149.5 mm **b** 150.5 mm **c** 1794 mm

4 a 35 g **b** 45 g **c** 180 g

5 a 100% **b** 33, 20, 22, 15, 11
c 101%, rounded values not accurate

6 a 800 **b** 16

7 a 7.55281818 **b** 7.6

8 7.5 cm, 8.5 cm **b** 5.5 cm, 6.5 cm
c 41.25 cm^2, 55.25 cm^2

9 166.375 cm^3

Pages 120–121 Standard form

1 a 8×10^{-6} **b** 6.7×10^{10}

2

Number	Rounded to 1 s.f.	Rounded to 2 s.f.	Rounded to 3 s.f.
0.004578	5×10^{-3}	4.6×10^{-3}	4.58×10^{-3}
34 640 000	3×10^7	3.5×10^7	3.46×10^7
0.00009638	1×10^{-4}	9.6×10^{-5}	9.64×10^{-5}

3 a 680 000 **b** 0.00089 **c** 985 000 000

4 a 3.45×10^7 **b** 7×10^4 **c** 2.1×10^9
d 8.4×10^3

5 a 8×10^6 **b** 1×10^{-3} **c** 1.6×10^4

6 a 1.1×10^{12} km^3 *(1 mark for 109...)*
b 3.6×10^8 km^2 *(1 mark for 3603...)*

7 1.2×10^{-10} m

8 a 3.53×10^{-2} cm **b** 1.24×10^{-3} cm^2
c 597.4 cm^2 (595.2 using rounded value in **b**)

Pages 122–123 Sequences

1 a 1 **b** 56 **c** $6n - 1$ **d** 8

2 –5

3 2, 5, 11, ... and –3, –5, –9, ...

4 1, 3, 5, ... *(1 mark for first two correct)*

5 a 1 **b** $3n$

6 $3n + 1$ *(1 mark for 3n)*

7 7

8 $4n + 4$ *(1 mark for 4n)*

Pages 124–125 Square numbers, primes and proof

1 1, 2, 3, 4, 6, 8, 12, 24
(1 mark for 6 correct)

2 8

3 36

4 36

5 121 or 144

6 a 1, 4, 9 *(1 mark for 2 correct)*
b 2, 3, 5, 7 *(1 mark for 3 correct)*
c 1, 2, 5, 10

7 6, 12, 18

8 1, 2, 4

9 a Either **b** Either **c** Even **d** Even

10 a 2 x any number is even.
b An even number plus 1 is odd.
c $2n(2m + 1) = 4nm + 2n = 2(2nm + n)$ which is a multiple of 2.

11 $n + n + 1 + n + 2 = 3n + 3 = 3(n + 1)$ which is a multiple of 3.
(1 mark for 3n + 3, 1 mark for justification)

Pages 126–127 Algebraic manipulation 1

1 a $a(b + c)$, $ab + ac$ and $(c - b) \div a$

2 a $8x - 12$ **b** $15a^2$ **c** $11a + 2b$
d $5x + 25$

3 a $n + 7$ **b** $n + 2$ **c** 6

4 a $5x + 3$ **b** $2y + 2$

5 a $5x - 4$ **b** $4x$ **c** $x - 4$ **d** $x + 1$

6 a $3 + x$, $3 + 2x$, $6 + 3x$
b $2x + 3$, $4x - 8$, $4x - 11$

7 a Perimeter = $6x + 12$
b Area = $2x^2 + 6x + 9$

Pages 128–129 Algebraic manipulation 2

1 a $12a^3b^2$ **b** $10a^5b^5$ **c** $8ab^2$ **d** $4a^3b^3$

2 a $x^2 + x - 12$ **b** $x^2 + 8x + 15$
c $x^2 - 3x + 2$ **d** $x^2 - 2x - 8$

3 a $2x^2 + 5x - 4x - 10 = 2x^2 + x - 10$
b i $2x^2 + 11x + 12$ **ii** $6x^2 + x - 1$
iii $12x^2 - 10x - 8$

4 a n^2, $5n$, $3n$, 15 **b** $n^2 + 8n + 15$
(1 mark for $n^2 + 5n + 3n + 15$)

5 a $x^2 - 9$ **b** $x^2 - 25$ **c** $x^2 - 1$
d $x^2 - 16$

6 $\frac{1}{2}(2p + 10)(p + 1) = \frac{1}{2}(2p^2 + 12p + 10) = p^2 + 6p + 5$

7 Area = $2x^2 + (x + 3)(2x + 5) = 2x^2 + 2x^2 + 11x + 15 = 4x^2 + 11x + 15$

Pages 130–131 Factorisation

1 $18ab$ and $4b$ **2** $5xy$ and $5xy^2$

3 $4(3a - 5)$ and $2(6a - 10)$

4 a $2xy(3xy + 12x^2)$ **b** $(x + 2)(x + 6)$

5 $2y(y^2 - 10)$, all the other expressions are the same.

6 a $5(x + 5)$ **b** $4x^2(3x - 1)$

7 a $ab(3a + 4)$ **b** $4a^2b^2(3a + b)$
c $2a^2b(3b^2 + 2)$ **d** $2a^2b^2(3b + 2a)$

8 a $(x - 4)(x + 3)$ **b** $(x - 5)(x - 3)$
c $(x + 1)(x + 2)$ **d** $(x + 6)(x - 4)$

9 a $(n + 2)$ and $(n + 5)$ **b** $(x + 4)(x - 1)$

Pages 132–133 Formulae

1 a Input 5, output 17 and Input 1, output 5

2 2.5 **3 a** 21 **b** 1

4 $3(x - 2)$ **5 a** £16 **b** £40

6 a 0.4 **b** 6 and 10

7 35 m *(1 mark for $a = 7.5\ ms^{-2}$)*

8 a Volume = 2360π cm^3, Area = 720π cm^2
b Volume = 1610 cm^3, Area = 795 cm^2

Pages 134–135 Graphs 1

1 a (7, 7) **b** (6, 7) **c** (21, 21)
d (20, 21)

2 (0, 8), (–2, 10) and (10, –2) **3** (2, 3)

4 $y = 5$, $x + y = 2$ and $x = -3$

5 a $x = -3$ **b** $y = x$ **c** $y = 4$
d $x + y = -2$

6 a i $x = 1$ **ii** $y = 3$ **iii** $x + y = -2$
b 18 square units

7 $y = x - 1$ **8** $y = 2x + 1$

Pages 136–137 Graphs 2

1 a $y = 2x + 1$ and $y = 2x - 3$ **b** $y = 2x + 1$ and $y = 4x + 1$

2 c, d, a, b **3** (3, 8) and (–2, –7)

4 $y = 2x - 2$

5

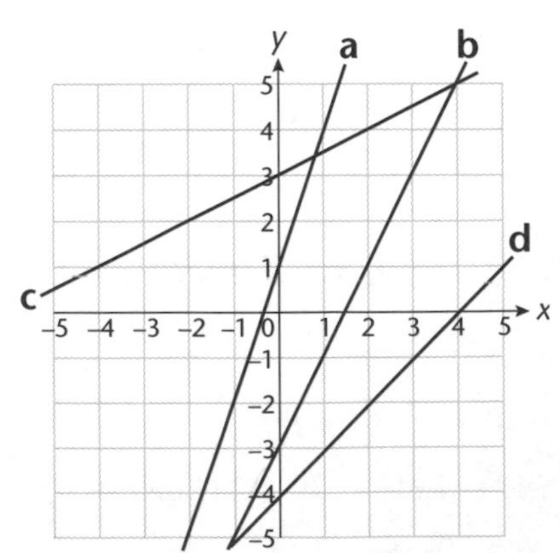

6 Graph C. A is unrealistic as items do not fall at a steady speed through the air. B is unrealistic as it implies that Jenny speeds up as she approaches the ground. D is unrealistic as it implies that Jenny does not descend at all.

7 a 6 km/h **b i** 3 km **ii** 10 min
c i 5 min **ii** 12 km/h

Pages 138–139 Equations 1

1 27 **2** 13.5

3 a 23 **b** 27

4 a 5 **b** 21

5 a 27 **b** 15 **c** 5.4 **d** 1.8

6 a 10.5 **b** 18 **c** 17.5 **d** 3

7 a 14 **b** –1

8 $(2x - 5) \times 4 = (x - 4) \times 2$

$$8x - 20 = 2x - 8$$
$$8x - 2x = -8 + 20$$
$$6x = 12$$
$$6x \div 6 = 12 \div 6$$
$$x = 2$$

(1 mark for 3 correct lines)

Pages 140–141 Equations 2

1 –10 **2** 6

3 a –7 **b** 2.5 **c** 2.7 **d** 2.25

4 a 1.5 **b** 2

5 a 2 **b** –5

6 a $x > -2$ **b** $-1 \leq x < 3$

7 $x < 14$

8 a $x \geq 4$ **b** $x > 9$

9 a $x > 2.5$ **b** $x \leq 0$

Pages 142–143 Trial and improvement

1 64

2 30

3 25–30

4 14

5 26.368

6 $x(x + 3) = 40$, sides are 5 cm and 8 cm and perimeter is 26 cm.

7 4.6 *(1 mark for testing a value between 4 and 5, 1 mark for testing 4.65)*

8 2.3 *(1 mark for testing a value between 2 and 3, 1 mark for testing 2.35)*

9 a Area = $x(x + 2) = 16.64$ **b** 3.2
(1 mark for testing value above 2)

Pages 144–145 Simultaneous equations

1 $y = 2x - 1$ and $2y + x = 8$

2 a $2(2x + 3) + x = 11$ gives $5x + 6 = 11$
b $x = 1$ **c** $x = 1, y = 5$

3 a $2(2x - 1) + 2(3y + 1) = 20$ gives $2x + 3y = 10$ and $2(3x) + 2(3y - 2) = 20$ gives $x + y = 4$
b $x = 2, y = 2$

4 $x = 3, y = 5$ **5 a** 70p **b** 50p

6 a $2x + 1 = 4y + 5$ *(1 mark)*, $2x - 4y = 4$ *(1 mark)*
b $2x + 1 + 4y + 5 + 3x + 2y = 20$ *(1 mark)*, $5x + 6y + 6 = 20$ *(1 mark)*
c $x = 2.5, y = 0.25$

7 a $2x + 2y = 14$, $3x + y = 16$
b $x = £4.50, y = £2.50$

Pages 146–147 Bearings

1 N 000° or 360°, NE 045°, E 090°, SE 135°, S 180°, SW 225°, W 270°, NW 315°

2

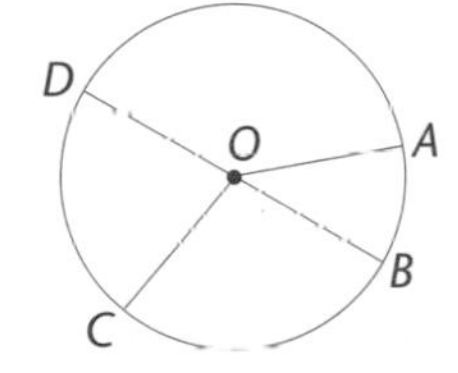

3 $a = 060°$ $b = 120°$ $c = 225°$

4 *A* is 9 km at 080°, *B* is 7 km at 230°, *C* is 10 km at 300°

5 *Shown half scale.*

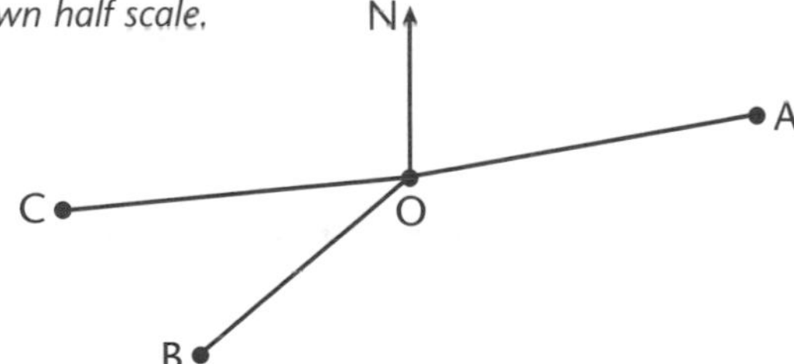

Pages 148–149 Angle facts

1 70° **2** 98°

3 87° **4** 103°

5 a 70° **b** 125°

6 Missing angle in triangle = $180 - (x + y) = 180 - z$

7 a Trapezium **b** 45°

8 168° **9 a** 108° **b** 36°

Pages 150–151 Angles in parallel lines and polygons

1 $a = 50°$ $b = 130°$ $c = 50°$

2 $d = 72°$ because alternate

3 $e = 55°$ because corresponding

4 $f = 120°$ because allied

5 $g = 50°$ because opposite, $h = 130°$ because allied

6 540°

7 Each interior angle is 120° and Each exterior angle is 60°.

8 $x = 72°$ $y = 108°$ $z = 72°$

9 Angles at any corner are 90° = 360° – 135° – 135° and all sides are equal.

Pages 152–153 Reflections and rotations

1 a 90° **b**

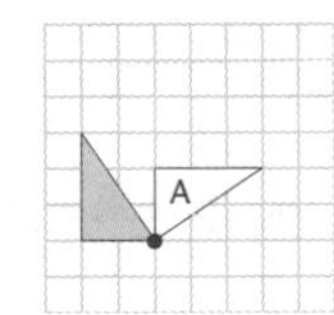

2

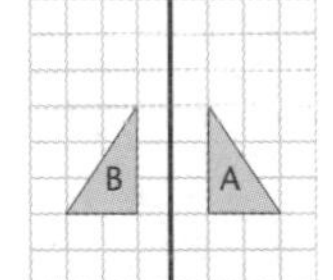

3 a Rotation *(1 mark)* of 180° *(1 mark)* about (0, 2) *(1 mark)*
b Reflection *(1 mark)* in the x-axis *(1 mark)*

4

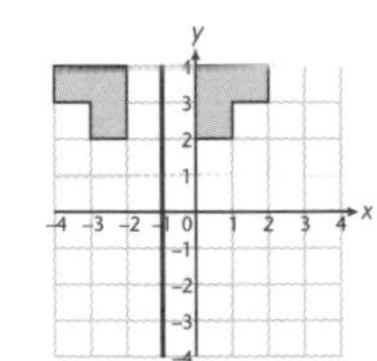

5

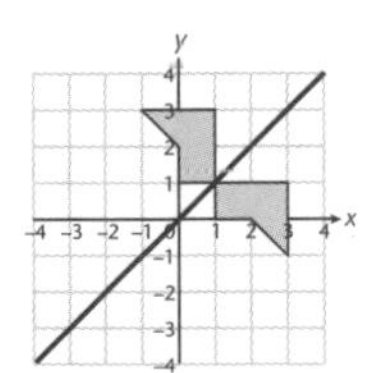

6

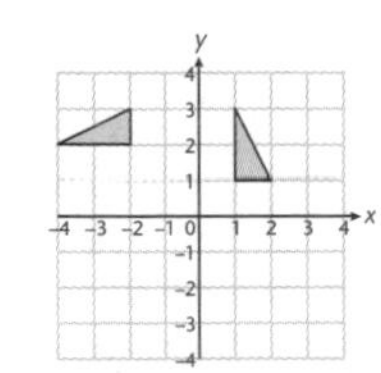

7 a Reflection *(1 mark)* in $x = 2$ *(1 mark)*
b Reflection *(1 mark)* in $y = -x$ *(1 mark)*
c Rotation *(1 mark)* clockwise through 90° *(1 mark)* about (0, –1) *(1 mark)*

Pages 154–155 Enlargements

1 a 2 **b** $2\frac{1}{2}$

2

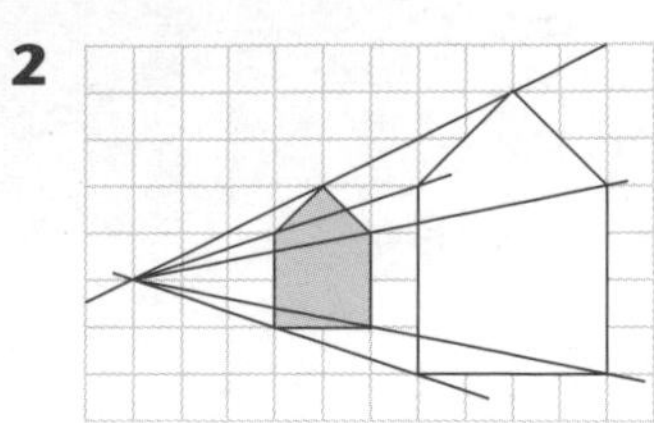

3

4

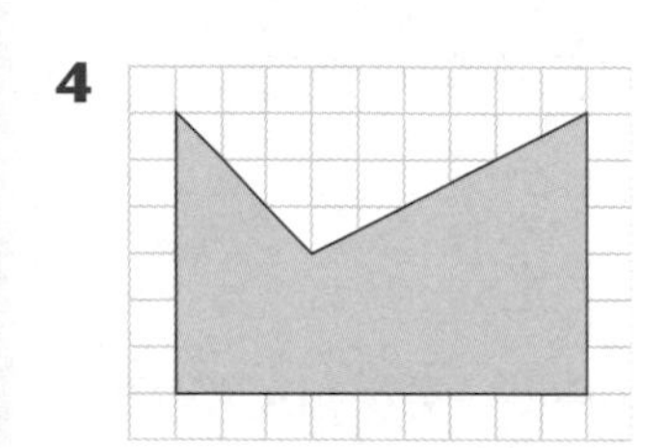

5 $A'(2, 2)$, $B'(2, 8)$, $C'(6, 2)$

6 A (2, 1), B (1, 2), C (2, 2)

7 a B **b** A **c** C

Pages 156–157 3-D shapes

1 Triangular prism

2 a, b and c

3 Square-based pyramid

4 a 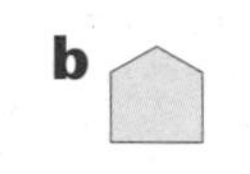 **b** 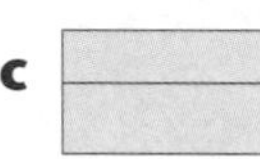 **c**

5 a 3 **b** Infinite number **c** 7

6 a, b

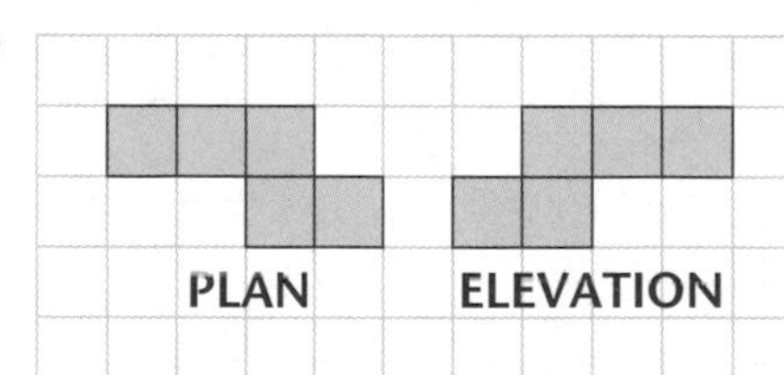

7

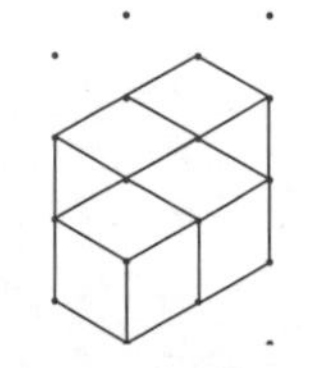

(1 mark for any isometric drawing with 4 cubes)

8 a ii $2\pi r^2 + 2\pi rh$

b i 1130 cm³ *(1 mark for $\pi r^2 h$)*

ii 603 cm² *(1 mark for 226 or 377)*

9 432 cm² *(1 mark for 216)*

Pages 158–159 Perimeter and area

1 a Perimeter = 12 cm **b** Area = 6 cm²
(1 mark for both units)

2 a Perimeter = 36 cm **b** Area = 60 cm²
(1 mark for both units)

3 A 6 cm², B 5 cm², C $4\frac{1}{2}$ cm², D 6 cm²

4 $7\frac{1}{2}$ cm² *(1 mark for units)*

5 21 cm² *(1 mark for units)*

6 a $x = 5$ cm **b** $y = 6.25$ cm

7 48 cm² *(1 mark for 36 cm² and 12 cm²)*

Pages 160–161 Circumference and area of a circle

1 15.7 cm **2** 25.1 m

3 8 cm **4** 24.6 cm *(1 mark for units)*

5 25.7 cm **6** 28.3 cm²

7 19.6 cm² *(1 mark for units)*

8 81π cm² **9** 50.3 cm²

10 21.5 cm²

Pages 162–163 Volume

1 a 15 cm³ **b** 46 cm²

2 2 cm **3** 210 cm² *(1 mark for units)*

4 4 cm **5** 8 cm **6** 6 m³

7 D = 180 cm³, C = 240 cm³,
A = 288 cm³, B = 625 cm³

8 800 l *(1 mark for 800 000 cm³ or 0.8 m³)*

9 4 cm

Pages 164–165 Constructions

1 *(1 mark for arcs, 1 for accuracy)*

2 *(1 mark for arcs, 1 for accuracy)*

3 *(1 mark for arcs, 1 for accuracy)*

4 *(1 mark for 2 correct sides, 1 mark for all correct)*

5 *(1 mark for 1 correct side and 1 angle, 1 mark for all correct)*

6 *(1 mark for 1 correct side and 1 angle, 1 mark for all correct)*

Pages 166–167 Loci

1 a 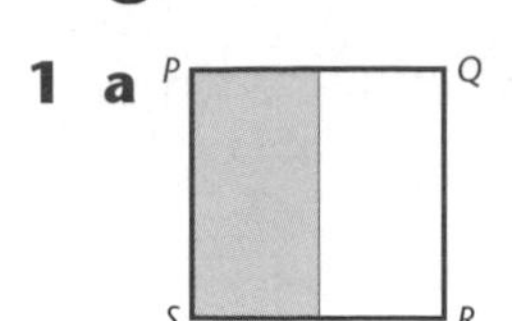

b 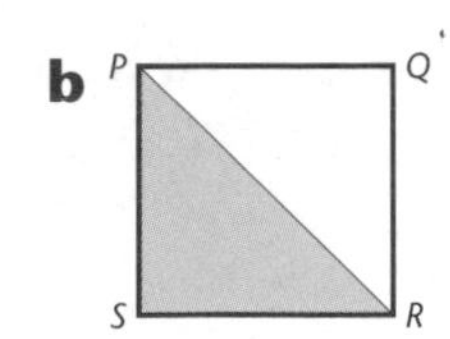

2 i b **ii** c **iii** a **iv** d

3 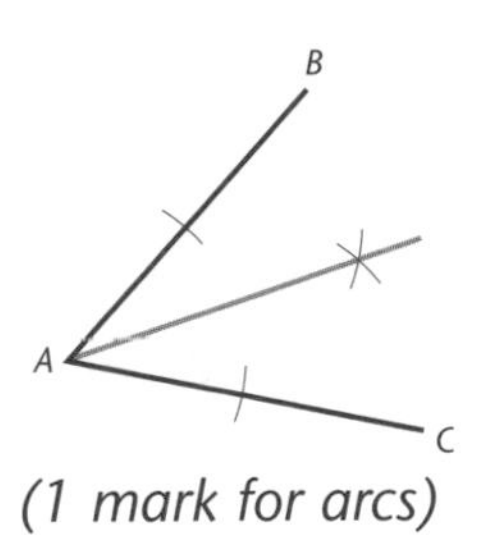

(1 mark for arcs)

4 a No

b *Shown half scale*

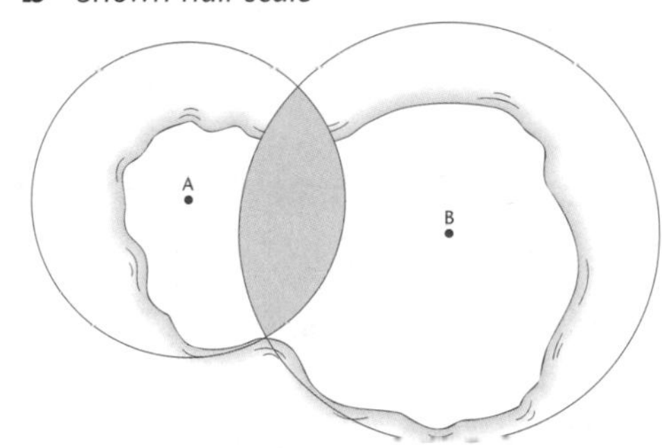

5 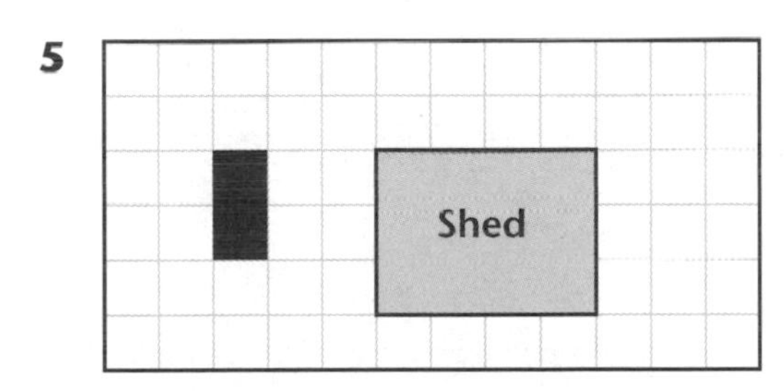

Pages 168–169 Similarity

1 a A and D as all sides the same
b A or D and C sides in same ratio

2 a 2 **b** 30°
c 8 cm
d 5 cm

3 a 2.5
b 50°
c 3 cm
d 6.25 cm

4 a All angles the same
b 3 cm
c 4.5 cm

5 a dimensions not in same ratio
b 6.67 cm

Pages 170–171 Pythagoras' theorem

1 a 25 cm² **b** 7.07 cm **c** 39.3 cm²

2 12.2 cm *(1 mark for $\sqrt{149}$)*

3 10.4 cm *(1 mark for $\sqrt{108}$)*

4 86.1 km *(1 mark for 36.1)*

5 10.2 km *(1 mark for $\sqrt{104.48}$)*

6 Yes, as $8^2 + 1.8^2 = 8.2^2$

7 14.1 cm *(1 mark for $\sqrt{200}$)*

Pages 172–173 Trigonometry

1 5.74 cm *(1 mark for 10 x sin 35)*

2 28.6° *(1 mark for $\tan^{-1}$ (6 ÷ 11))*

3 15.9 cm *(1 mark for 12 ÷ cos 41)*

4 30° *(1 mark for $\sin^{-1}$ (6 ÷ 12))*

5 236° *(1 mark for $\tan^{-1}$ (30 ÷ 20) or 56°)*

6 120.6° *(1 mark for $\tan^{-1}$ (5.2 ÷ 8.8) or 30.6°)*

7 a 7.42 cm *(1 mark for $\sqrt{55}$)*
b 68° *(1 mark for $\cos^{-1}$ (3 ÷ 8))*
c 22.2 cm² (or 22.3 cm² from rounded value for height) *(1 mark for $\frac{1}{2} \times 6 \times h$)*

Pages 174–175 Sectors and circle theorems

1 2 units *(1 mark for $2\pi r = \pi r^2$)*

2 a 19.5 cm *(1 mark for 7.5)*
b 22.6 cm² *(1 mark for $\pi \times 6^2 \div 5$)*

3 a i $\frac{1}{4}$ **ii** $\frac{2}{9}$
b A is 34.2 cm², B is 28.3 cm²
c A is 23.8 cm, B is 21.4 cm

4 a B is 1570.8 cm², A is 1295.9 cm²
b B is 4712.4 cm³, A is 3534.3 cm³

5 $a = 18°$, $b = 90°$, $c = 32°$, $d = 52°$

6 $a = 82°$, $b = 25°$, $c = 140°$, $d = 142°$

Pages 176–177 Comparing distributions

1 a C **b** M **c** M **d** C

2 a Mode: 10X: 0, 10Y: 4
b Median: 10X: 2, 10Y: 2.5
c Mean: 10X: 2.2, 10Y: 2.5
d 10Y: as bigger averages

3 B, as more consistent and mode is 0

4 a, b and c

5 a both 2.5 **b** Aisha 7, Betty 3
c Aisha, as she sometimes scores lots of goals or Betty, as she is more consistent

6 a i 7 **ii** 7 **b i** 10 **ii** 4
c Outside is more consistent or greenhouse has more tomatoes on some plants

Pages 178–179 Line graphs

1 a 14 °C **b** Thursday and Saturday
c Friday 17 °C
d Line has no meaning. Temperature changes throughout the day. Values are just the values at 12 midday.

2 a Midnight 3 °C, Midday 6 °C
b 6 °C
c Friday 12 °C

3 a 1900 miles **b** 5500 miles
c June

4 a 9000–10 000
b No, doesn't necessarily drop below 5000
c Yes, change is continuous, about 15 000

5 No, can't assume it keeps raining but if graph continues at same rate it will flood by 2 pm.

Pages 180–181 Statistical diagrams

1 Distributions are same but men are taller by about 10 cm

2

0	7 8 8 8 9
1	1 2 3 3 5 7 8 9
2	1 2

3 a 34 years **b** 22 years **c** 21 years

4 a Angles: Blue 162°, White 72°, Silver 126°
b

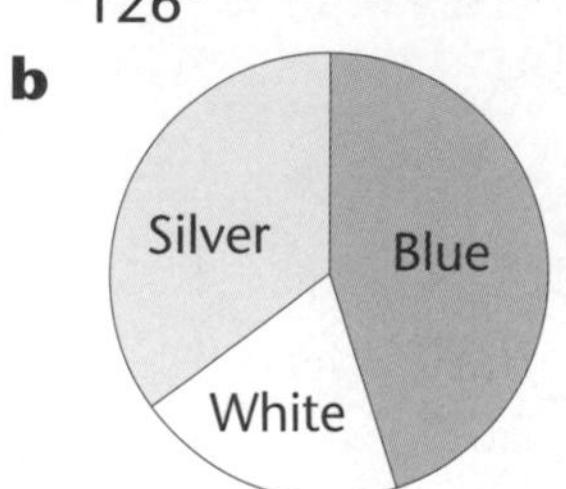

5 160 **6** 40 **7** c **8** a

9 Angles: British 162°, American 90°, French 72°, German 36°

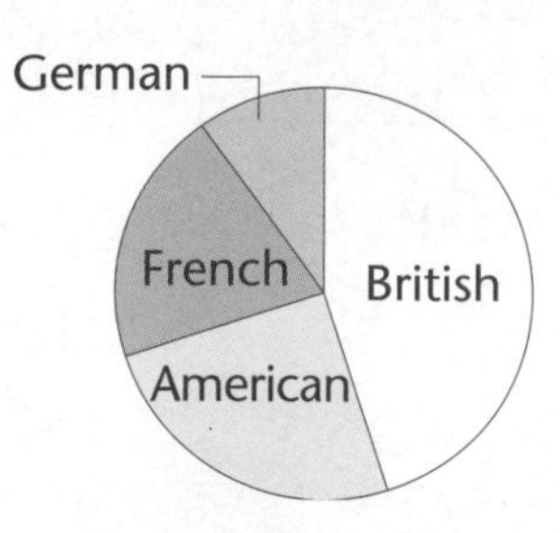

10 Angles: Blue 216°, Silver 108°, Black 36°

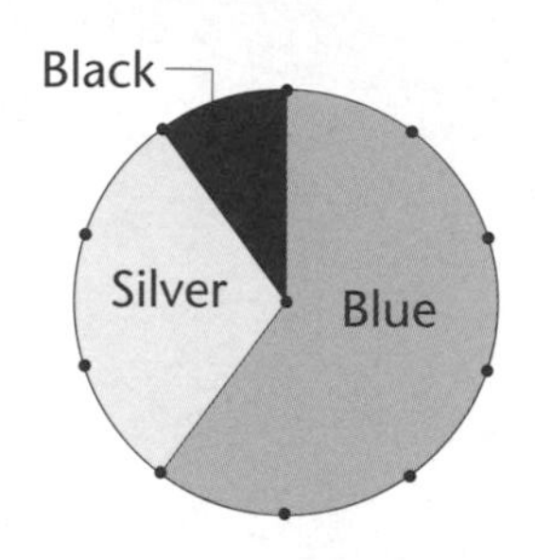

Pages 182–183 Scatter diagrams

1 a i W **ii** N **iii** K **iv** G
b i N **ii** G **iii** S **iv** K

2 a

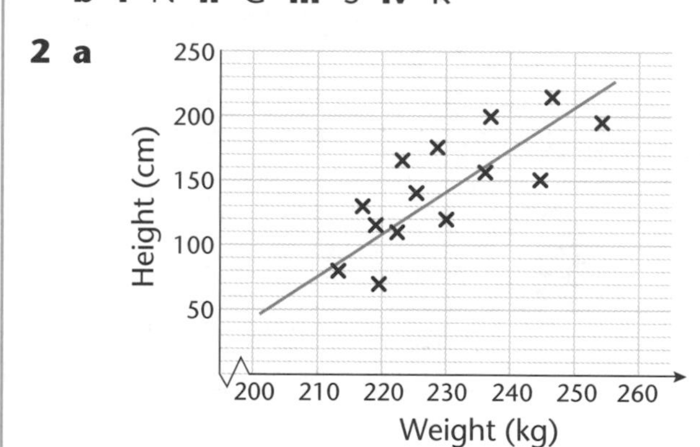

b 150 cm
c No, it doesn't fit line of best fit.

3 2 hours **4** a, c and d

5 a, b and d

6 a Weak positive
b

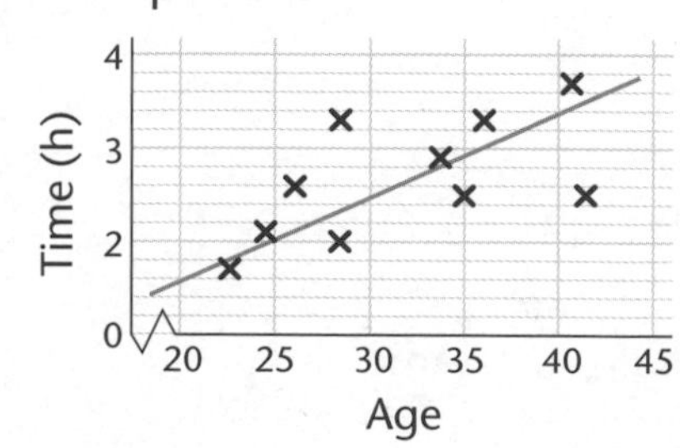

c $3\frac{1}{2}$–4 hours

Pages 184–185 Surveys

1 b and c

2 **a** **i** Not enough responses
ii Not enough responses
iii Not relevant to question
b Good range of responses that covers all views

3 1 Leading question, 2 offensive question

4 Derek. He gets a random sample while others are biased.

5 Not enough responses

6 Quota sampling, will give a varied sample

7 All the men and only a fraction of women is not representative.

8 **a** **i** Not enough responses
ii Not enough responses
iii Not relevant to question
b Good range of responses which can be analysed mathematically

9 Owen. All methods give a random sample but the more students surveyed the better.

Pages 186–187 Box plots and cumulative frequency diagrams

1 **a**

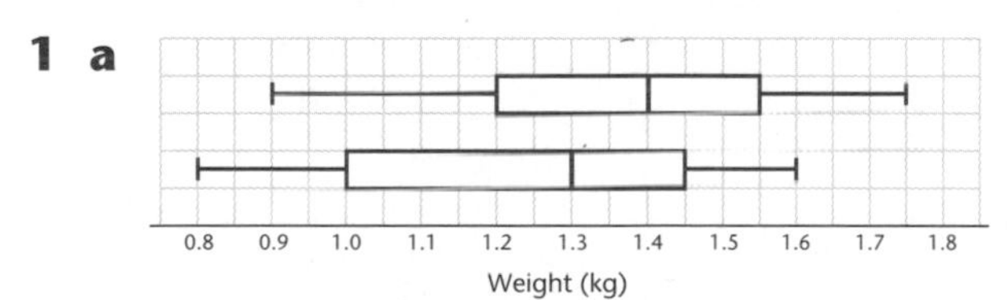

b The marrows grown inside are 0.1 kg heavier on average.
c The marrows grown outside are less consistent.

2 **a** The upper and lower quartiles and lowest and highest values are symmetrical about the median.

b

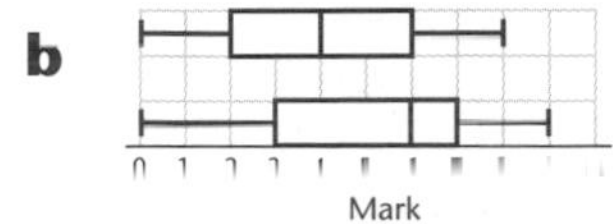

(1 mark for 3 parts correct)

3 **a** **i** 141 cm **ii** 10 cm **b** 8

4 **a** IQR: 14 cm, median: 40 cm
b IQR: 19 cm, median: 52 cm
c Trees in park are bigger by about 12 cm and also they are more consistent with a lower range.

Pages 188–189 Probability 1

1 $\frac{3}{7}$ **2** 3 blue balls **3** $\frac{11}{20}$

4

0 — C — A — B — $\frac{1}{2}$ — D — 1

(1 mark for 3 correct values)

5 **a** $\frac{2}{5}$ **b** $\frac{2}{3}$

6 **a** $\frac{2}{5}$ **b** $\frac{2}{3}$

7 Red 8, Blue 2

8 **a** $\frac{1}{4}$ **b** $\frac{1}{2}$

9 **a** $\frac{1}{6}$ **b** $\frac{1}{12}$ **c** $\frac{1}{6}$

Pages 190–191 Probability 2

1 **a** **i** 0 **ii** 1 **b** **i** $\frac{1}{4}$ **ii** $\frac{3}{4}$

2 **a** Probabilities are 0.4 and 0.6 on each pair of branches.
b **i** 0.16 *(1 mark for 0.4 x 0.4)*
ii 0.36 *(1 mark for 0.6 x 0.6)*

3 **a** $\frac{1}{3}$ *(1 mark for* $\frac{6}{10} \times \frac{5}{9}$*)*
b $\frac{7}{15}$ *(1 mark for* $\frac{1}{3} + \frac{2}{15}$*)*

4 Bag B as $\frac{4}{7}$ is bigger than $\frac{5}{9}$
(1 mark for either probability)

5 **a** 40 **b** $\frac{1}{625}$ **c** $\frac{48}{625}$

6 **a** **i** $\frac{1}{6}$ **ii** $\frac{5}{6}$
b **i** $\frac{1}{6} \times \frac{5}{6} + \frac{5}{6} \times \frac{1}{6} = \frac{10}{36}$
ii He loses £1.60

7 **a** Tom, as most throws
b 6 white and 14 black, $320 \div 1000 \approx 0.3$ and $0.3 \times 20 = 6$

Pages 192–193 Practice Paper 1 (non-calculator)

Question	Mark	Correct response	Comments
1 a	1	For example,	Acute angles are less than 90° and obtuse angles are between 90° and 180°.
b	1	An acute angle is less than 90° and four times a number less that 90° must be less than 360.	This is a Using and Applying maths question. In your answer, you must show that you know the sum of the angles in a quadrilateral is 360°.
2 a	1	$\frac{12}{25}$	Your answers to this question must be written as a fraction. Answers such as 12 out of 25 or 12 in 25 or 12 : 25 are not acceptable.
b	1	$\frac{8}{25}$	
c	1	$\frac{20}{25}$ or $\frac{4}{5}$	20 chocolates are not white. You would not lose the mark if you did not cancel down the fraction.
3	2 or 1	(£)3.75 digits 3375 seen	Use a suitable method to work out 1.35 x 25, which is 33.75. The saving is 33.5 – 30 = 3.75
4 a	1 1	18 19	Work out 6 + (2 x 5) + 2 = 6 + 10 +2 = 18 Work out (3 x 6) + 5 – (2 x 2) = 18 + 5 – 4 = 19
b	1	7	$a + b + c = 13$, so $d = 20 - 13 = 7$
5 a	2	80, 60%	Remember that a percentage means out of 100.
b	1	Any that work: 5 out of 100, 1 out of 20, etc.	
6	1 1 1	$\frac{11}{12}$ $\frac{7}{12}$ $\frac{4}{12}$ or $\frac{1}{3}$	$\frac{1}{2} = \frac{6}{12}$ $\frac{1}{4} = \frac{3}{12}$ and $\frac{1}{3} = \frac{4}{12}$ $\frac{3}{4} = \frac{9}{12}$
7 a	1	x: 2, 4, 6 y: 7, **9**, **11**	The mapping $y = x + 5$ means add 5 to each x-value to get the y-value.
b	1	x: 2, 4, 6 y: 1, **5**, **9**	The mapping $y = 2x - 3$ means multiply each x-value by 2 and then subtract 3 to get the y-value.
c	1	$(y) = \frac{1}{2}x + 1$ or $(y) = x \div 2 + 1$	To get the y-value, you halve each x-value and then add 1.
8 a	2 or 1	Cuboid B	The surface area of a cuboid is the total area of its 6 faces. A = 62 cm^2, B = 88 cm^2, C = 82 cm^2. You would get 1 mark for finding the correct surface area for two cuboids.

Question	Mark	Correct response	Comments
b	2 or 1	Cuboid C	The volume of a cuboid is $V = lwh$. A = 30 cm^3, B = 40 cm^3, C = 42 cm^3. You would get 1 mark for finding the correct volume for two cuboids.
c	1	1 (cm)	$V = 10 \times 3 \times h$, so $30 = 30h$ and $h = 1$
9 a	1	No. The sides are the same length, but the 4 angles are not 90°, or it is a rhombus.	
b	1	Yes. Two pairs of adjacent sides have the same length.	
c	1	No. It only has one pair of parallel sides, or it is a trapezium.	
10	1 1 1	–2 –4 –5	12 + (–2) = 10, since +(–) is the same as – 6 – (–4) = 10, since –(–) is the same as + –2 x (–5) = 10, since – x – = +
11	2 or 1	$\frac{1}{4}$ $\frac{6}{24}$ or $\frac{3}{12}$	To multiply fractions, multiply the numerators and the denominators. You would get 1 mark for not cancelling.
12 a	1	(–3, –2)	Draw the line AB to find the mid-point.
b	1	(–1, 0)	
13	1 1 1	$(x) = 4$ $(y) = 5$ $(z) = 3$	$2x = 8$ (take 3 from both sides) $x = 4$ (divide both sides by 2) $3y - 6 = 9$ (multiply out brackets) $3y = 15$ (add 6 to both sides) $y = 5$ (divide both sides by 3) $2z - 4 = 2$ (take z from both sides) $2z = 6$ (add 4 to both sides) $z = 3$ (divide both sides by 2)
14	1 1	$x = 7 - y$ $w = z \div 3$ or $\frac{z}{3}$	Take y from both sides to make x the subject. Divide both sides by 3 to make w the subject.
15	2 or 1	$x^2 - x - 12$ most of $x^2 + 3x - 4x - 12$ seen	You would get 1 mark for 3 terms from $x^2 + 3x - 4x - 12$
16 a	1	Positive correlation or greater height, greater arm span	

Question	Mark	Correct response	Comments
b	1	Suitable line, for example,	Your line should have about the same number of points above it as below. It may or may not pass through some of the points.
c	1	8 cm	Use your line of best fit.
d	1	No, outside range of graph and babies may not follow same growth pattern as older students.	
17 a	1	95 (kg), 12	Interpret the final entry in the diagram for the heaviest person. Count the number of entries in the last 3 rows for those over 70 kg.
b	1	83 (kg), 6	Those over 70 kg are in the last 2 rows.
c	1	76 (kg), 30 (kg)	The modal weight is the weight common to most members. The range is the difference between the heaviest weight and the lightest weight.
18 a	2 or 1	891 660 and/or 81 seen	Either work out 33 x 20 + 33 x 7 = 660 + 231 or 27 x 30 + 27 x 3 = 810 + 81
b	1 1	3. The smallest 2 digit number is 10 and 10 x 10 = 100 4. The largest 2 digit number is 99 and 99 x 99 = 9801	
19 a	1	0.36	Work out 36 ÷ 100
b	1	0.58	Work out 290 ÷ 500
c	1	Zara, because she had more trials	
20	1 1	2.4×10^6 6×10^5	When multiplying, you add the powers. When dividing, you subtract the powers.
21	1 1 1	$a = 64$, $b = 1$ $a = 4$, $b = 3$ $a = 2$, $b = 6$	Start by thinking of all the factors of 64. These give possible values for a.

Pages 199–205 Practice Paper 2 (calculator)

Question	Mark	Correct response	Comments
1 a	2 or 1	30 (people) 12 seen	84° is equivalent to 7 people, so 84 ÷ 7 = 12° is equivalent to 1 person. 360 ÷ 12 = 30
b	2 or 1	162 (degrees) 18 seen	20 people in a pie chart will get 360 ÷ 20 = 18° per person. 9 people will be an angle of 9 x 18 = 162°
2 a	1 1 1	12 or 16 15 17	The factors of 48 are: {1, 2, 3, 4, 6, 8, 12, 16, 24, 48}. You can give both answers. The factors of 150 are: {1, 2, 3, 5, 6, 10, 15, 25, 30, 50, 75, 150}. 'Between' means that you do not include 10 or 20. The factors of 51 are: {1, 3, 17, 51}.
b	1	'No' ticked and an explanation such as '150 is not in the 60 times table.'	You need to make it clear that you understand that a multiple is in the times tables so writing down 60, 120, 180, … would just about do this.
3 a	1	12	The range is the difference between the highest and the lowest numbers. From –3 to 9 is a difference of 12.
b	1	20	The total of the negative numbers is –5. The total of the positive numbers is 25. 25 – 5 = 20
c	1	–1	The mode is the most common number.
d	1	1	The median is the middle number when the numbers are in order. These are already in order but there is an even number of values, so the median is between 0 and 2.
e	2 or 1	2.5 Showing a correct method, e.g. the total ÷ 8	The mean is the total of the numbers divided by how many numbers there are. The total is 20 and there are 8 values.
4 a	1	23 or 37 or 43 or 47	Prime numbers have no factors other than 1 and themselves. Only one answer is needed but you will not lose the mark if you give more than one.
b	1	25 or 36	Square numbers are numbers that can be written as 5 x 5 or 6 x 6, etc.
c	1	Because a square number always has a factor other than 1 or itself.	You need to make it clear you know that square numbers can be written as a product such as 2 x 2, 5 x 5, etc.
5 a	1	75 (degrees)	As the triangle is isosceles, the two base angles are the same. 180 – 30 = 150, 150 ÷ 2 = 75
b	1	135 (degrees)	There are 360° in the full turn. The total of the angles shown is 45 + 90 + 90 = 225. 360 – 225 = 135

Question	Mark	Correct response	Comments
6 a	1	110 (degrees)	Be careful to choose the correct scale on your protractor.
b	2 or 1	350 (metres) 7 cm seen	Multiply the length of BC by 50.
7	3 or 2 or 1	(£)62.50 125 seen 1000 grams seen	This is a Using and Applying maths question. You have to convert 1 kg to grams (1000 grams), then divide 1000 by 8 (= 125). You then have to change 125 fifty pence coins into pounds.
8 a	2 or 1	Any three points on the line Any two points	The possible values are: (–2, –1), (–1, 0), (0, 1), (1, 2), (2, 3), (3, 4), (4, 5), (5, 6). You can read the coordinates from the graph.
b	1	$y = x + 1$	You should see that the second (y) coordinate is equal to 1 more than the first (x) coordinate.
c	1	A line parallel to $y = x + 1$ passing through (0, 3)	The line is parallel to the given line but passes through 3 on the y-axis rather than 1.
9 a	1	219.8 to 220 (cm)	The formula for the circumference is $C = \pi d$ or $C = 2\pi r$.
b	2 or 1	2200–2300 digits 22 or 23 seen	5 kilometres is 5000 metres which is 500 000 cm. 500 000 ÷ (π x 70) = 2273.64. The answer only has to be approximate, so you can round off.
10	1 1	1694 3.5	Remember to include the bracket keys. Work out the numerator and denominator separately first.
11 a	1	29.4 (cm)	Work out 21 x 1.4
b	1	164 (mm)	Work out 229 ÷ 1.4
c	1	Yes, folded paper is 210 mm x 147 mm	Work out 294 ÷ 2 and compare widths and heights.
12	2	Man City by 57 seats (Man Utd 2295, Man City 2352)	Work out 3.4 ÷ 100 x 67 500 and 4.9 ÷ 100 x 48 000
13	1	Yes, 3 x (2 x 18.5 + 13) = 150, 76 + 4 x 18.5 = 150	Substitute x = 18.5 in each side of the equation.
14	1 1 1 1	$16 - 10a$ $b^2 + 4b$ $24c^2$ $4d^2$	Collect like terms.
15	2 or 1	1.9 total of 38 seen	For each row, multiply the number of children per family by the frequency. Add these values to give 38 and divide by 20.

Question	Mark	Correct response	Comments
16 a	1	135°	Sum of angles in an octagon = 6 x 180 = 1080°, so each angle is 1080 ÷ 8 = 135
b	1	45°	b = 360 – 135 – 90 – 90 = 45
17	2	1 mark for arcs. 1 mark for angle 60° ± 1° 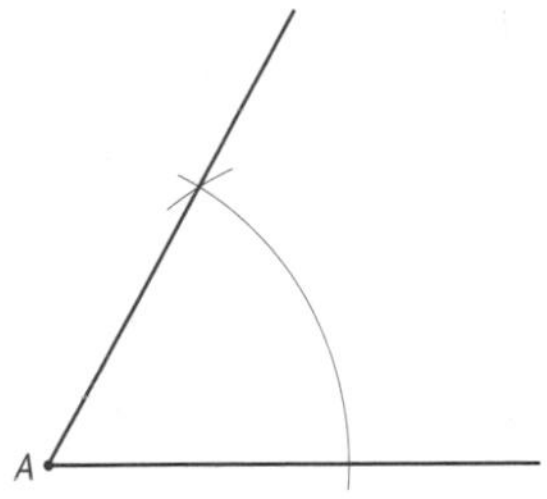	Think how to construct an equilateral triangle using a ruler and compasses only.
18 a	2 or 1	8.22 (cm) $\sqrt{67.6}$ seen	Use Pythagoras' theorem. $x^2 = 5.4^2 + 6.2^2$
b	2 or 1	4.34 (cm) 9.3 x tan 25 seen	Use trigonometry. Tan 25 = $\frac{\text{opp}}{\text{adj}} = \frac{x}{9.3}$
19	2 or 1	Increases by 8% 108 or 0.9 x 1.2 seen	Length = 90% of original Width = 120% of original Area = 0.9 x 1.2 of original
20	2 or 1	$3n + 1$ $3n$ seen	Each term is one more than 3 times the term number.
21	2 or 1	20 cm AF = 12 cm	Using similar triangles $\frac{ED}{EC} = \frac{EF}{EA}$ so $\frac{4}{10} = \frac{8}{20}$

Pages 206–207 Mental Mathematics Test

Each question is worth 1 mark each, giving you a total out of 30.

Question	Correct response
1	43
2	3
3	14
4	–10
5	31
6	0.9
7	95 (94.9 recurring)
8	Friday
9	Circle
10	$(3\frac{1}{2}, 2)$
11	10
12	4.5
13	$\frac{17}{20}$
14	33.3% (33%)
15	36
16	21
17	475
18	0.5 x 0.3 x 0.2
19	10
20	7
21	12 or 16
22	199
23	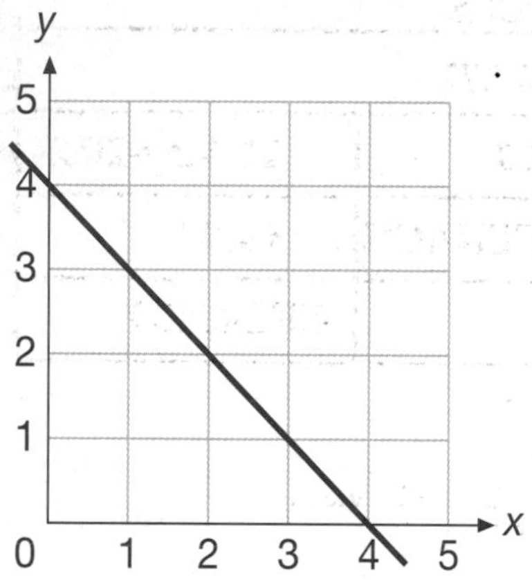
24	9π
25	12
26	375–425
27	$x - 4$
28	$\frac{1}{3}$
29	Tetrahedron
30	120